Using Activities

from the "Mathematics Teacher" to Support *Principles and Standards*

Edited by

Kimberley Wilke Girard
Glasgow High School
Glasgow, Montana

Margaret Plouvier Aukshun
Billings West High School
Billings, Montana

NATIONAL COUNCIL OF
TEACHERS OF MATHEMATICS

ISBN 0-87353-566-9

Using Activities

from the "Mathematics Teacher" to Support *Principles and Standards*

Introduction

The "Activities" section has been a regular feature of the *Mathematics Teacher* since 1972. Looking at the topics represented by the articles gives a snapshot of what and how mathematics was being taught during the past thirty years. One example is that calculators were primarily used for computation in the 1970s, microcomputer programs in BASIC and Logo were used to investigate polynomial graphs and geometric properties in the 1980s, and then graphing calculators were used to explore fractal geometry in the 1990s. In 2012, what will we be seeing in the "Activities" section from the 2000s?

In selecting activities for this volume, we looked for those that would help teachers as they reach toward the vision of mathematics teaching and learning described in NCTM's *Principles and Standards for School Mathematics*. Space limitations prohibited the inclusion of all the activities printed in the last decade. The quality and variety of the work made it difficult to narrow our original "wish list." In the end, we chose the activities that we, as classroom teachers, have been using. They represent topics ranging from slope and volume to nonperiodic tilings and the methods of voting. Some activities may replace a lesson in a textbook, others may supplement a lesson, and still others may introduce the students to applications of mathematics not previously considered. Notes from the authors include objectives, materials needed, suggestions for using the activities, and sometimes extensions of the activity.

The grid on the next page will help teachers choose those activities that best meet the needs of their students. Grade levels suggested by the authors of the activities are included. The Standards from *Principles and Standards* are identified as applicable to the activity in a major way (in dark blocks) as well as to a lesser extent (in lighter blocks). Since many activities contain elements of more than one Standard, they have been arranged by the order of their appearance in the *Mathematics Teacher* rather than by topic.

Biographical information about authors, which is included with each activity, was current at the time of original publication and may no longer be accurate.

		Standards										
Activity Name	Page Number	Number & Operation	Algebra	Geometry	Measurement	Data Analysis & Probability	Problem Solving	Reasoning & Proof	Communication	Connections	Representation	**Grade Levels Suggested by Authors**
Playing with Blocks: Visualizing Functions	3			■				■		■		9–12
Generating and Analyzing Data	13		■			■					▨	11–12
The Functions of a Toy Balloon	21		■		▨							7-12
Graphing Art	27		■								▨	9–12
Time for Trigonometry	35			▨			▨					9–12
What Is the r For?	45	▨				■	▨					9–12
The Inverse of a Function	53		■								▨	9–12
Experiments from Psychology and Neurology	63		▨			■						9–12
Packing Them In	71			■			▨		▨			8–12
More Functions of a Toy Balloon	79		■	▨	▨					▨		9–12
Activities for the Logistic Growth Model	87		■									9–12
The Secret of Anamorphic Art	97			■						▨		9–12
Probability, Matrices, and Bugs in Trees	107	▨				■						10–12
Forest Fires, Oil Spills, and Fractal Geometry: Part 1	117					■				■	▨	9–12
Nonperiodic Tilings: The Irrational Numbers of the Tiling World	129			■			▨					7–12
Forest Fires, Oil Spills, and Fractal Geometry: Part 2	139		■	▨								9–12
Calculating Human Horsepower	147		■		■							7–10
Graphing for All Students	151		■								▨	7–10
Whelk-come to Mathematics	159		■			■						10–12
Discovering an Optimal Property of the Median	173						■	■				8–12
Will the Best Candidate Win?	185						▨			■		8–14
The Jurassic Classroom	193		▨	■								8–11
Volume of a Pyramid: Low-Tech and High-Tech Approaches	203			■								8–12
A Stimulating Study of Map Projections	211	▨								■		7–12
Sports and Distance-Rate-Time	223			▨	■							8–10
Print Shop Paper Cutting: Ratios in Algebra	231				■		▨			▨		7–12

Playing with Blocks: Visualizing Functions

Miriam A. Leiva, Joan Ferrini-Mundy, and Loren P. Johnson

November 1992

Edited by Karen A. Dotseth, *Cedar Falls High School, Cedar Falls, Iowa;* Kim Girard, *Nashua High School, Nashua, Montana;* Mally Moody, *Oxford High School, Oxford, Alabama*

- In grades 9–12, the mathematics curriculum should include the continued study of the geometry of two and three dimensions so that all students can interpret and draw three-dimensional objects. (NCTM 1989, 157)

- The teacher of mathematics, in order to enhance discourse, should encourage and accept the use of . . . concrete materials used as models. (NCTM 1991, 52)

- The teacher of mathematics should pose tasks that . . . engage students' intellect; develop students' mathematical understanding and skills; stimulate students to make connections and develop a coherent framework for mathematical ideas. (NCTM 1991, 25)

Introduction: The NCTM's *Curriculum and Evaluation Standards for School Mathematics* (1989) argues for "increased attention on the analysis of three-dimensional figures" (p. 158), as well as for the importance of the function concept as a major unifying idea in mathematics. In the spirit of promoting connections among mathematical topics and helping students to form multiple representations of mathematical situations, this "Activities" develops open-ended investigations that combine spatial-visualization activities and emphasize aspects of the function concept. Problems dealing with maxima and minima in a physical setting create a foundation for more abstract work.

These investigations are designed to engage secondary school students in a noncontrived, nontrivial use of manipulatives, in writing about mathematics, in making conjectures and validating them through inductive processes, and in working with partners and in cooperative groups. The authors acknowledge here an assessment task furnished by Jan de Lange that served as the springboard for this activity. The *Middle Grades Mathematics Project* (Lappan et al. 1986) is a good resource for related activities.

Grade levels: 9–12

Materials: A set of worksheets, many centimeter blocks, centimeter grid paper, rulers. Blocks of any size will do, but the use of centimeter blocks on the grid paper is convenient.

Objectives: Students will make connections among the concrete, pictorial, and abstract representations of a three-dimensional figure and will investigate relations and functions by—

- representing a structure from pictorial representations;

- building one or more structures that correspond to a given pictorial representation, map, or number of blocks;

- recognizing the rules that relate a structure to its pictorial representation, its map, and the numbers of blocks required to build it; and

- exploring these rules as either relations or functions and determining if a function is onto or one-to-one.

Prerequisites: Students should be able to read instructions and complete tables; no knowledge of algebra is expected.

Directions: The setting for all the problems is a 4-cm-by-4-cm-by-2-cm frame.

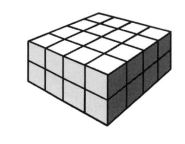

Mariam Leiva teaches at the University of North Carolina at Charlotte, Charlotte, North Carolina. She is the editor for the K–6 Addenda Series. Joan Ferrini-Mundy is interested in research in the learning of calculus, the professional development of teachers, and gender-related issues in mathematics. She teaches at the University of New Hampshire, Durham, New Hampshire. Loren Johnson has had extensive experience as a mathematics educator and is currently a doctoral student at the University of New Hampshire, Durham, New Hampshire.

Students begin by building structures with centimeter cubes and recording various views of the structure, focusing on top view, front view, and side view. Unless otherwise specified, the top view of a structure has in each of the 4-cm-by-4-cm squares a number that denotes the numbers of blocks placed on that location of the structure. Thus, a top view is a map, or representation, of the structure and is a blueprint to make congruent structures. For this activity, however, front view and side views are not numbered. Instead, the 2-cm-by-4-cm views, or maps, are shaded to represent the presence of a block or blocks on the corresponding row in the first or second level of the structure.

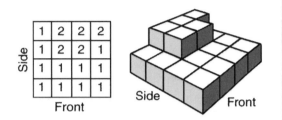

For example, the top view illustrated corresponds to the front view and side view shown here.

Front view   Side view

In the activities the assumption is made that views are given from a left-to-right orientation, that is, a reflection of a view may yield a different structure. Through the problems and class discussions, the teacher will lead students to discover that a numbered top view yields a unique structure, whereas the front view and side view may represent one or more structures.

For most of the activities, students work in pairs or small groups; they could also work individually. Either way, encourage whole-class discussions and instruct students to reflect on their conjectures and answers to justify them to the class. Each of the four sheets of activities requires approximately one class period to complete. Sheet 4 could easily be extended to two class periods, especially if students are encouraged to look for generalizations to determine minimum block requirements.

Sheet 1 introduces the idea of a buildable structure and supplies students with methods for recording various views of the structure. The activity emphasizes spatial orientation and visualization, methods of thinking that are important for many areas of mathematics translation between actual structures and other representations for those structures. In exploring the relationships between front views and back views and between different side views, the ideas of symmetry and reflection can conveniently be raised. Note that task 3 requires the students to work in a group and then in task 4 they work in pairs. In tasks 4 and 5, students compare two pairs of structures so each pair of students considers two examples.

Sheet 2 helps students reach the conclusion that a front-view–side-view pair does not describe a unique structure, although a top-view representation does. The concept of a relation that maps a front-view–side-view pair to a structure or a top view to a structure might naturally be discussed with algebra students in conjunction with this activity. The issue of uniqueness of the image in this relation also should emerge, and the distinction between function and relation can be made.

Sheet 3 extends the activities of sheet 2 by introducing questions about the numbers of blocks involved in building various structures. Ideas of maximum and minimum are introduced in a very practical setting. Questions about the minimum number of blocks are likely to challenge students, particularly in reconsidering what information is given in a front view or a side view; the discovery that the front view or side view doesn't furnish any record of "depth" is solidified in this activity. In addition, relations mapping front-view–side-view pairs or top views to the maximum or minimum number of blocks involved are introduced. Discussion of when these relations are functions would be appropriate at this point. In addition, notions of domain, range, image, one-to-oneness, and ontoness will arise. (A function $f{:}A \to B$ is *onto* if, for any $b \in B$, an $a \in A$ exists such that $f(a) = b$.)

Sheet 4 focuses on the functional relationship between front-view–side-view pairs and the minimum number of blocks possible for making the structure. Question 2 illustrates that the function fails to be one-to-one, and question 3 promotes exploration of the possibility that the function is onto. Investigation of the maximum number can also be explored in the context of sheet 4. These activities set the stage for the function concept, and the teacher may wish to encourage students to discuss and generalize the ideas that arise in this connection.

Answers: Sheet 1

1. *a*) Yes
   b) Back view:

In this instance, the back view and the front view are the same, although they will not always be so.

c) Other side view:

Open spaces are flipped (reflected) from right to left (reflected horizontally). The observer should visualize the view by facing it.

2. *a*) Drawings will vary.
   b) Drawings will vary.
   The open (and shaded) blocks are flipped from right to left. Yes, it will always hold.

   c) The views are horizontal (left to right) reflections of each other.
   Yes. Imagine a vertical axis drawn in the middle of a side view. Flipping the side view about the axis produces the other side view.

3. Teachers may want to regulate the movement of students on this problem. Drawings will vary.

4. No, not usually. The probability that the same structure will occur is small, since usually many possibilities arise. The number of different structures depends on the top view.

5. Yes. Only one structure can be built to match the top view.

Sheet 2

1. *a*)

All students should have the same structures. The top view is the blueprint for a congruent structure.

b) Top views will vary. No, structures will differ. Front and side views do not fix the positions of blocks.

2. *a*) No. It's not possible to have a side view with blocks stacked zero or one high in each position when the front view has blocks stacked two high.

b)

Answers will vary.

---

c) If either view has a height of two blocks in any position and the other view has no height of two anywhere, then the structure is not buildable.

3. Yes. An example:

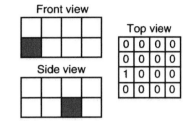

4. Only one. A top view specifies the exact number of blocks in each position.

5. No. The only positions that might be questionable would be those that contain a "1," but it is not possible to put a block specified by a "1" in the second level. Thus no unbuildable top views are possible.

Sheet 3

1. Examples (others are possible);

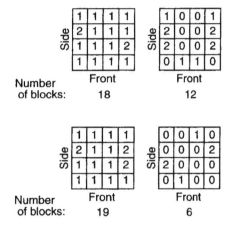

2. *a*) 20
   b)

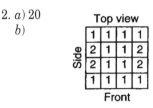

   c) Other top views include the following:

0	0	1	0
2	0	0	0
0	0	0	2
0	1	0	0

0	1	0	0
0	0	0	2
2	0	0	0
0	0	1	0

0	1	0	0
2	0	0	0
0	0	0	2
0	0	1	0

3. a) 6

b)

Top view

0	0	1	0
0	0	0	2
2	0	0	0
0	1	0	0

c) Here are three possibilities (answers will vary):

0	0	1	0
2	0	0	0
0	0	0	2
0	1	0	0

0	1	0	0
0	0	0	2
2	0	0	0
0	0	1	0

0	1	0	0
2	0	0	0
0	0	0	2
0	0	1	0

When viewed from the front, four ways are possible of locating each double stack as one progresses from front to back in the structure and similarly for each single stack.

4. Block arrangements using from six to twenty blocks are possible. Several possibilities are illustrated:

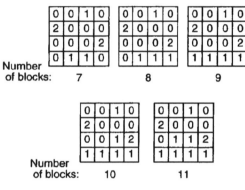

Number of blocks: 7   8   9

0	0	1	0
2	0	0	0
0	0	1	2
1	1	1	1

0	0	1	0
2	0	0	0
0	1	1	2
1	1	1	1

Number of blocks: 10   11

Sheet 4

1. Minimum number of blocks: 4, 6, 6, 10, 8, 1. To find the minimum number of blocks, count the number of stacks of 2 and the number of stacks of 1 in the front view and the side view, then sum the number of blocks in the largest number of stacks of 2 and the number of blocks in the largest number of stacks of 1.

2. Here are two examples:

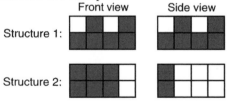

Front view   Side view

Structure 1:

Structure 2:

3. 1; 32

4. No. No structure is possible of a side view and a front view having a minimum greater than 11. Answers may vary, but some examples are shown in figure 1.

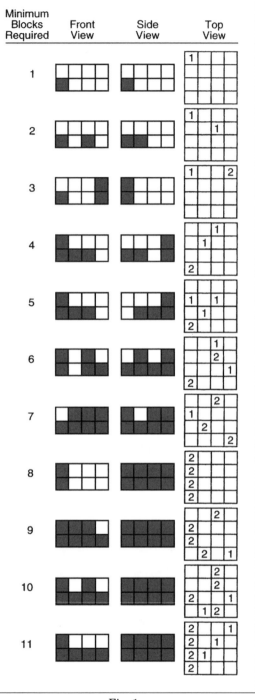

Fig. 1

REFERENCES

Lappan, Glenda, Elizabeth Phillips, William Fitzgerald, Janet Schroyer, and Mary Winter. *Middle Grades Mathematics Project: Spatial Visualization.* Menlo Park, Calif.: Addison-Wesley Publishing Co., 1986.

National Council of Teachers of Mathematics. *Curriculum and Evaluation Standards for School Mathematics.* Reston, Va.: The Council, 1989.

———. *Professional Standards for Teaching Mathematics.* Reston, Va.: The Council, 1991.

PLAYING WITH BLOCKS

The problems on this sheet refer to structures
built within a 4-cm-by-4-cm-by-2-cm "frame." All
pictures of structures must be "buildable" in the
sense that they could actually be created with
blocks.

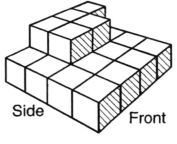

Side   Front

You should work independently on tasks 1 and 2. Your teacher will assign you to a group for
task 3, and you will be paired with another student for tasks 4 and 5.

1. Examine the top view and the corresponding front view and a side view of a structure
given below. The numbers on the squares of
the top view represent
the number of blocks
that are on that loca-
tion. The shaded
squares on the front
view and side view
show the location of
blocks on the corre-
sponding row and level
of the structure.

Top view

Side
1	1	2	1
1	2	2	0
0	2	2	0
1	1	1	1

Front

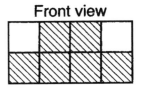

Side view   Front view

a) With your blocks, make the structure described by the top view. Does the front view of
our structure correspond to the front view given? _____

b) Draw the back view of the structure. How does the back
view relate to the front view?

_____

_____

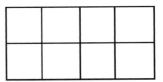

c) Look at the side views of the structure. One is given above;
draw the other one. How are the side views related?

_____

_____

2. Build a structure of blocks.
   a) Record the top view.

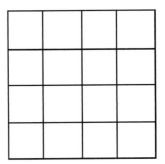

From the *Mathematics Teacher*, November 1992

b) Draw the front and back views. How does the back view relate to the front view? _____

_____

Front view

Back view

Do you think this relation always holds? _____

_____

_____

Side view 1

Side view 2

c) Draw and label both side views. How are the two sides related? _____

_____

_____

If one side view of a structure is given, could you draw the other side view? _____

Explain. _____

_____

d) Try other structures to check your conjectures in parts (*b*) and (*c*).

3. Build another structure. Draw and label the front view and side view on grid paper and turn it in. Your teacher will give you another group's paper. Walk around the classroom and find the structure that matches your sheet.

4. Build a structure within your 4-cm-by-4-cm-by-2-cm "frame" so that your partner cannot see it. On your grid paper, draw and label the front and side views. Then swap papers with your partner and build a structure using your partner's "blueprints." (Don't tear down your structure.) Next compare your structures. Are they the same? _____ How many different structures do you think can be built to match the views? _____

5. Repeat exercise 4, but draw the top view instead of the front and side views. Are the structures congruent? _____ How many different structures can be built to match the top view? _____

The problems on this page refer to structures built within a 4-cm-by-4-cm-by-2-cm "frame."
All pictures of structures must be buildable unless otherwise indicated; a structure must
include at least one block. Work with a partner, and use blocks.

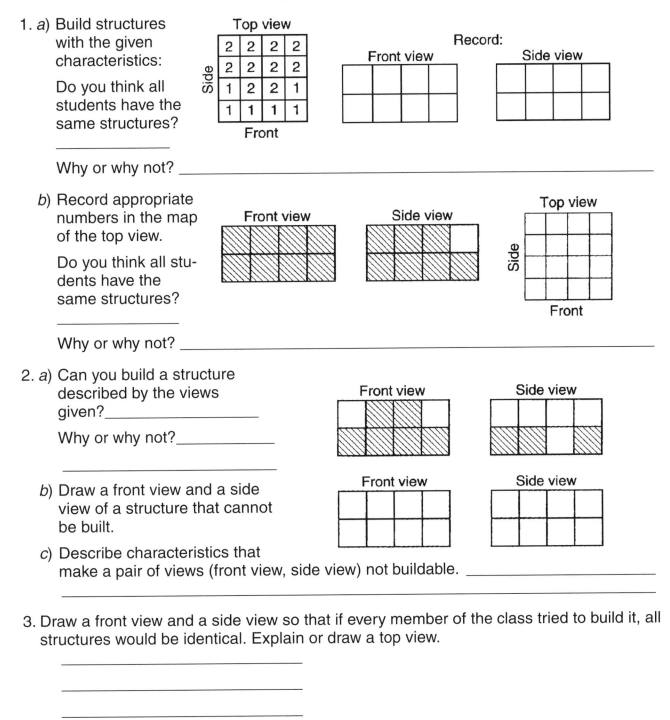

1. *a*) Build structures
    with the given
    characteristics:

    Do you think all
    students have the
    same structures?

    _____

    Why or why not? _____

    b) Record appropriate
    numbers in the map
    of the top view.

    Do you think all stu-
    dents have the
    same structures?

    _____

    Why or why not? _____

2. *a*) Can you build a structure
    described by the views
    given?_____

    Why or why not?_____

    _____

    b) Draw a front view and a side
    view of a structure that cannot
    be built.

    c) Describe characteristics that
    make a pair of views (front view, side view) not buildable. _____

    _____

3. Draw a front view and a side view so that if every member of the class tried to build it, all
   structures would be identical. Explain or draw a top view.

    _____

    _____

    _____

4. If you are given a top view of a structure, how many different structures can you make?

    _____ Why? _____

5. Can you draw a top view of a structure that is not buildable? _____
   Explain on the back of this sheet.

From the *Mathematics Teacher*, November 1992

Use the front and side views given here
for the following problems.

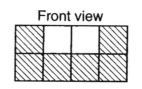

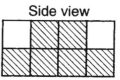

1. Build as many structures as you can that have these views. Record each top view and
   the number of blocks necessary to make the structure.

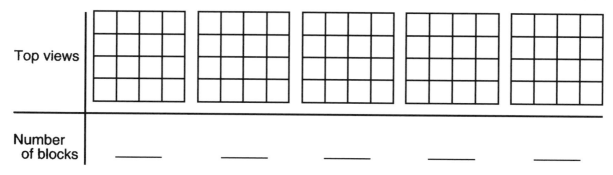

Top views

Number of blocks _____   _____   _____   _____   _____

2. a) What is the largest (maximum) number of blocks you can use to build a structure having
      these views? _____

   b) Record the top
      view of a struc-
      ture using the
      maximum num-
      bers of blocks.

   Top view

   Side

   Front

   c) Draw a different
      top view using
      the maximum
      number of
      blocks.

3. a) What is the smallest number of blocks (minimum) you can use to build the structure
      having these views? _____

   b) Record the top view of a structure using the
      minimum number of blocks.

   c) Draw a different top view using the minimum
      number of blocks.

4. For each integer between the maximum and minimum in problems 2 and 3, try to build
   with that many blocks a structure having the given front and side views. For each possible structure, draw the top view. Use the back of this sheet if necessary.

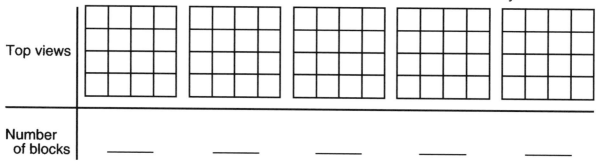

Top views

Number of blocks _____   _____   _____   _____   _____

Given a front-view–side-view pair, determine the minimum number of blocks with which it is possible to make a buildable structure having those views. A structure must include at least one block.

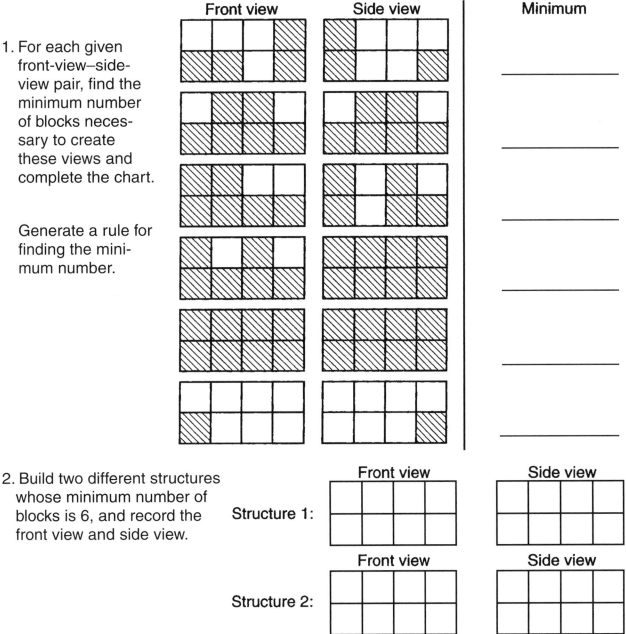

Front view	Side view	Minimum

1. For each given front-view–side-view pair, find the minimum number of blocks necessary to create these views and complete the chart.

   Generate a rule for finding the minimum number.

2. Build two different structures whose minimum number of blocks is 6, and record the front view and side view.

   Structure 1:   Front view   Side view

   Structure 2:   Front view   Side view

3. How many blocks are in the *smallest* possible structure that can be made in a 4-cm-by-4-cm-by-2-cm frame? _____
   How many blocks are in the *largest* possible structure? _____

4. Is it possible for each of the integers between 1 and 32 to be a minimum number for some front-view–side-view pair? Determine the validity of your conjecture by building these structures. Be prepared to discuss and defend your conclusions in class.

Generating and Analyzing Data

Jill Stevens

Edited by Karen A. Dotseth, *Cedar Falls High School, Cedar Falls, Iowa;* Kim Girard, *Nashua High School, Nashua, Montana;* Mally Moody, *Oxford High School, Oxford, Alabama*

September 1993

TEACHER'S GUIDE

Introduction: The Curriculum and Evaluation Standards for School Mathematics (NCTM 1989) calls for the continued study of functions so that all students can—

- model real-world phenomena with a variety of functions;

- represent and analyze relationships using tables, verbal rules, equations, and graphs;

- translate among tabular, symbolic, and graphical representations of functions;

- recognize that a variety of problem situations can be modeled by the same type of function;

- analyze the effects of parameter changes on the graphs of functions. (NCTM 1989, 154)

The curriculum standards also stress the continued study of data analysis and statistics so that all students can—

- construct and draw inferences from charts, tables, and graphs that summarize data from real-world situations;

- use curve fitting to predict from data. (NCTM 1989, 167)

so that all students can...use curve fitting to predict from data.

The following activities promote these objectives.

Grade levels: 11–12

Materials: Activity sheets; M&M's; paper plates, cups, or other containers; stopwatches; birthday candles; matches; heat-resistant tiles or plates; rulers; centimeter grid paper; weather data; metersticks; jug or coffee pot with a spigot; scissors; graphing calculators

Objectives: Students will learn to analyze data to determine the type of function that most closely matches the data. They should demonstrate an understanding of how modifying parameters will change the graphs of functions by writing equations for those functions.

Prerequisities: Students should understand the relationships between various functions and their graphs. Their knowledge should cover linear, quadratic, higher-degree-polynomial, exponential, and trigonometric functions. They should also be able to use a graphing calculator with features that will allow them to edit data, create scatterplots, and draw functions over those plots.

Directions: The activities will require three to five class periods to complete with students working in groups of three to four. The following problem will introduce the activities and allow time for any necessary instruction on using a graphing calculator. One explanatory example is usually sufficient for students successfully to use a graphing calculator for these activities.

As an example, our class discussed a unit called "How Does Corn Grow?" to familiarize students with data and graphing. Students planted some corn. After the seeds sprouted, they measured their growth each day. Analyze the collected data.

Day 1 is the first day that sprouts appeared. Each group chose a sprout and measured its height for five days. The following results from one group are shown.

Day	Height (cm)
1	1.50
2	2.94
3	3.91
4	6.02
5	7.11

Ask students to enter the data into a graphing calculator, make a scatterplot of the data on the calculator, and analyze the results.

Before making the scatterplot, students must decide on an appropriate domain and range so that all their points will appear on the graph. In this example, we have chosen the day number as the

Jill Stevens teaches and serves as department chair at Irving High School, Irving, Texas. Her current interests include the College Board's Pacesetter program and using technology to enhance the understanding of mathematics.

independent variable (x) and the height of the corn plant as the dependent variable (y). We will let the x-axis go from -5 to 10 and the y-axis go from -5 to 15. In so doing, all data will appear on the screen so that both axes can be seen. After setting the appropriate domain and range, students should use the calculator to produce a scatterplot of the data and analyze the results.

Students should decide what type of function the data most closely represents and write an equation to fit the data. In this example, the data appear to be linear. The students can make an "eyeball" fit of the data by examining the scatterplot and determining a y-intercept and slope. They might try a y-intercept of 0.05 and a slope of 1.3. The students should use the calculator to draw the equation $y = 1.3x + 0.05$ over the scatterplot to check the accurracy of the fit, which is shown in figure 1. The equation appears to fit the data closely. In many examples, students will try several equations before they find a close fit.

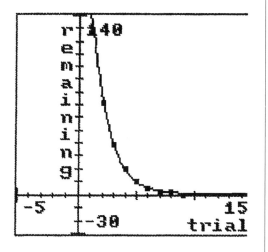

Fig. 1. Graph and equation from "How Does Corn Grow?" example; x scl = 0.5, y scl = 1.

Allow time for students to explain why they chose a particular function

Allow time for students to explain why they chose a particular type of function, how they derived their equation, and what restrictions the problem situation places on the domain and range of the function. In our example, x and y can be any real number in the algebraic equation, but in the model, x must be a positive integer and y must be a positive real number. Also, an upper bound will be necessary for x and y, since the corn will not continue to grow indefinitely.

The students should also discuss how their equation could be used to predict future function

values. They could use their equation to predict the height of a corn sprout on future days. If they had actually grown the corn plants, they could test their predictions by later measuring the specific sprouts.

The same general procedure that was outlined in the example problem should be followed for the activity sheets. The student should collect data, produce a scatterplot, analyze it, choose a function, and write an equation. Test the equations by drawing the function over the scatterplot on the graphing calculator. Class discussion should include why a certain function type was chosen, whether another function type might also work, how equations were derived, what restrictions are placed on the domain and range by the problem situation, and whether the model could be used to predict future values. Although no one answer is correct for each problem, some answers may be better than others. Some sample solutions are given.

Answers: Sheet 1: Figure 2 graphs the following data for "EliM&Mination" and sample solution 1.

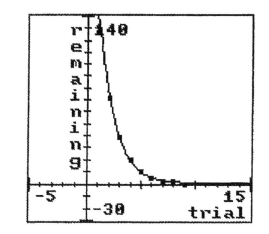

Fig. 2. M&M's activity graph; $y = 300\,(0.5)^x$. x scl = 1, y scl = 10.

Trial Number	Number Remaining
1	126
2	72
3	39
4	20
5	10
6	4
7	1
8	1
9	0

Figure 3 gives another view of data from sample solution 2.

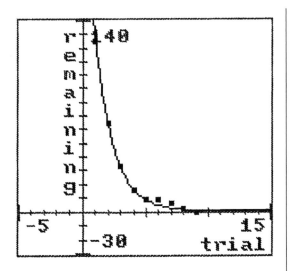

Fig. 3. Another solution to M&M's problem;
$y = (0.5)^{(x-8)}$.
x scl = 1, y scl = 10.

Trial Number	Number Remaining
1	125
2	65
3	33
4	15
5	9
6	8
7	6
8	2
9	0

A possible solution to "The Flaming Function" example graphs the data presented here (see fig. 4).

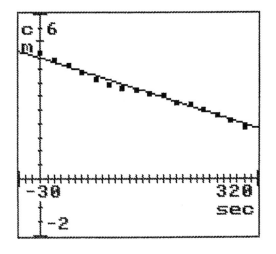

Fig. 4. "The Flaming Function's" possible solution;
$y = (-1/125)x + 4.4$.
x scl = 10, y scl = 0.5.

Time (sec)	Height (cm)
0	4.6
20	4.3
40	4.1
60	3.9
80	3.6
100	3.4
120	3.3
140	3.2
160	3.1
180	3.0
200	2.8
220	2.7
240	2.5
260	2.3
280	2.1
300	1.9

Sheet 2: Figure 5 graphs these data for "All Boxed In" sample solution.

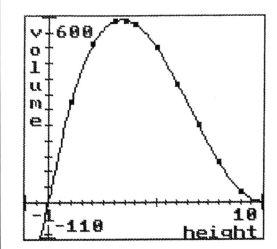

Fig. 5. "All Boxed In" problem is graphed with
$y = x(20 - 2x)^2$.
x scl = 0.5, y scl = 50.

Height (cm)	Volume (cm^3)
1	324
2	512
3	588
3.5	591.4
4	576
5	500
6	384
7	252
8	128
9	36

" 'Weather' It's a Function" was calculated from the normal high temperature for the Dallas area and was obtained by writing to a local television weather forecaster; see figure 6. The average monthly temperature for many cities can be found in almanacs. This information reduces the domain to twelve values but still yields a fairly smooth curve. Be sure the calculator is in the radian mode to work this problem.

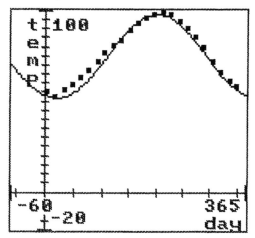

Fig. 6. " 'Weather' It's a Function" graph of a trigonometric
sine wave;
$y = 23 \sin(0.017214x - 115) + 75$.
x scl = 50, y scl = 5.

Day	Temperature
1	55
15	53
32	56
46	59
60	63
74	67
91	72
105	77
121	81
135	84
152	89
196	98
213	99
227	98
244	94
258	90
274	85
288	80
305	72
319	66
335	61
349	58

"Water Level" calculations are presented here. See the resulting graph in figure 7.

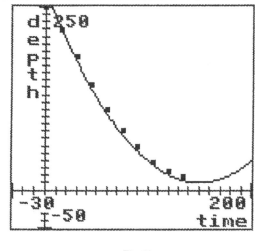

Fig. 7

Time (sec)	Depth (mm)
0	249
15	219
30	182
45	144
60	110
75	80
90	58
105	36
120	24
135	17

Discussion and extension activities: The "EliM&Mination" problem serves as a model of decay. Have students suggest real-life situations that this problem might model.

The "Flaming Function" is linear. It is interesting to compare the equation the students obtain from the "eyeball" method with those from the median-median-line and linear-least-square methods of curve fitting. The Data Analysis software package from the National Council of Teachers of Mathematics (1988) is a good source for information about these methods.

"All Boxed In" presents a cubic model. This problem could be extended by asking students to find the length plus girth and the surface area of each box in addition to its volume. The problem would then include a linear, quadratic, and cubic model. The post office uses the concept of girth to limit the size of boxes sent in the mail (length + girth ≤ 108 inches). Ask students to find the box with the greatest volume for the least surface area that would meet postal regulations.

The problem "Weather' It's a Function" produces a sine wave. The students can use this model to predict temperatures and then check to see how close their predictions come to the actual numbers.

The "Water Level" problem yields data for a quadratic model. An interesting activity would be to have different groups use different-sized jugs and compare the curves obtained from the data.

If computers are available, students may want to print their graphs rather than draw them. The figures illustrated in this article were created using The Mathematics Exploration Toolkit (IBM 1988). Another source of software is Data Analysis (NCTM 1988), which allows students to print tables and graphs. The software also includes many algorithms to fit curves to data sets.

REFERENCES

International Business Machines Corporation. The Mathematics Exploration Toolkit. Boca Raton, Fla.: The Corporation, 1988. Software.

National Council of Teachers of Mathematics. *Curriculum and Evaluation Standards for School Mathematics.* Reston, Va.: The Council, 1989.

North Carolina School of Science and Mathematics (NCSSM). Data Analysis. Reston, Va.: National Council of Teachers of Mathematics, 1988. Software.

The following activities will allow you to collect sets of data. As these activities are completed, record the data in tabular form.

Pour a half-pound bag of M&M's onto a paper plate so that the candies are one layer thick. You will need to spread the M&M's to the edges of the plate. Remove all the M&M's with the M showing on one side (look closely at the yellow ones because the M is hard to see). Count and record the number of M&M's removed and the number remaining. *Eliminate* the M&M's removed and pour the ones remaining into a container. Shake the container and pour these M&M's back onto the plate and again remove all the M&M's with the M showing. Record the number removed and the number remaining. Continue to repeat this process until all the M&M's are removed. Use the following chart to record your information. Add additional trial numbers as the experiment progresses.

Trial Number	Number Removed	Number Remaining
1		
2		
3		
4		
5		

Let x be the trial number and let y be the number of pieces remaining. Plot all points (x, y) and analyze the data. Make a scatterplot of each set of data on a graphing calculator and decide which type of function best represents the data. Write an equation that fits the data as closely as possible. Test your equations by drawing the function over the scatterplot on the graphing calculator. Though no *one* correct answer exists for each problem, some answers may be better than others. Try to find the best fit possible. Record the type of function you chose, the equation for the function, and a graph of the scatterplot and function.

THE FLAMING FUNCTION

For this activity, use the smallest-sized candles available and don't place the ruler too close to the flame.

Let x be time in seconds and y be height in centimeters. Let the initial value of x be zero and the initial value of y be the height of a birthday candle. Stand the candle on a heat-resistant tile or plate by lighting another candle and dripping some wax onto the plate and then setting the candle in the wax. Light the birthday candle and measure its height every twenty seconds. Extinguish the candle before it burns all the way down. Plot the ordered pairs (x, y) and analyze the data.

Take a 20-cm-by-20-cm piece of grid paper and cut congruent squares from each corner. Fold up the sides to form a rectangular-shaped box with no lid. Determine the volume of the box and complete the following chart. Repeat this process on other 20-cm-by-20-cm pieces of grid paper for several different-sized choices of cut squares.

Height (cm)	Length	Width	Volume (cm³)

Let x be the height of the box and let y be the volume of the box. Plot the ordered pairs (x, y) and analyze the data.

"WEATHER" IT'S A FUNCTION
Most years have 365 days. The first day of the calendar year is 1 January; 15 January is the fifteenth day of the calendar year; 1 February is the thirty-second day of the calendar year, and so on. Decide whether a relationship exists between the number of the day of the year and the normal high temperature for that day for the first and fifteenth days of each month. Use weather data from your local area. Let x be the number of the day of the year and y be the normal high temperature on that day. Plot all ordered pairs (x, y) and analyze the data. The following chart format can be used.

Day	Temperature

WATER LEVEL
Fill a gallon jug or a coffeepot with a spigot with water. Measure the depth of the water in millimeters. Turn on the spigot for fifteen seconds and measure the depth of the water again. Repeat this process until the water is at the level of the spigot. Let x be the total time the spigot was turned on in seconds (s) and let y be the depth of the water in millimeters. Plot the ordered pairs (x, y) and analyze the data using the following chart format.

Time (s)	Depth (mm)

The Functions of a Toy Balloon

Loring Coes III   November 1994

Edited by Timothy V. Craine, *Central Connecticut State University, New Britain, Connecticut;* Kim Girard, *Glasgow High School, Glasgow, Montana;* Guy R. Mauldin, *Science Hill High School, Johnson City, Tennessee*

TEACHER'S GUIDE

Introduction: Real data can break down the barriers between geometry and algebra and can bring these topics vividly to life. When students collect and examine their own data, the lessons they learn are all the more convincing. The following activity should grab students' attention and show some dramatic connections between the *algebraic* concepts of slope, linear functions, and power functions and the *geometric* concepts of circle, sphere, volume, and π.

Grade levels: 7–12

Materials: Activity sheets; high-quality, spherical party balloons, 12 inches in diameter; stopwatch; bow calipers, purchased from a science supply outlet or borrowed from the science teacher; centimeter measuring tapes; calculator or computer program capable of curve fitting, such as the TI-82 or CA-Cricket Graph III

Objectives: Students will gain experience measuring with different tools: stopwatch, centimeter tape, and bow calipers. They will learn about the uses and limitations of different ways of measuring. They will explore different graphs and make connections between algebra and geometry. Students will learn to understand, interpret, and appreciate discrepancies in data. Advanced students will become familiar with curve-fitting computer programs and precise mathematical models.

Prerequisites: The activity on sheet 1 should be accessible to all students in grades 7–12. Sheet 2 is appropriate for students learning about slope. Sheet 3 can be used by students in grades 9–12 who have access to computer software that can handle regressions.

Directions: Assign students to small groups. Each group needs one student in each of the following roles: a balloon inflator, a caliper user, a tape-measure user, a timer, and a recorder. Give each group one balloon and a copy of sheet 1. Groups can share calipers and measuring tapes, if necessary. The recorder enters on the activity sheet all the data generated by the group.

Sheet 1

The balloon inflator blows up the balloon as much as possible with only one breath. The inflator then firmly pinches the mouthpiece to prevent the balloon from deflating but does not tie off the mouthpiece. While the inflator holds the balloon, another student measures the balloon at its widest diameter with the calipers. The student with the centimeter tape first measures the gap in the calipers and then measures the circumference of the balloon at its widest point. If calipers are unavailable, the balloon can be placed on the floor against a wall and a book placed on the floor on the other side of the balloon. Then the distance from the wall to the book can be measured. The balloon inflator releases the balloon. The timer uses the stopwatch to measure how long the balloon flies, from the moment of release until the balloon "dies," not until it hits the ground. Each group repeats the activity, adding breaths one at a time until the inflator feels that the balloon will burst.

Sheet 2

Real data appeal to students of all ages. Students can use the data generated on sheet 1 to complete the graphs on sheet 2. Several graphs are possible: diameter versus circumference, number of breaths versus diameter, breaths versus flight time, and so on. Each type of graph offers a different function for discussion. Middle-grade students can determine that some of these relationships appear

Loring (Terry) Coes III is chair of the mathematics department at Rocky Hill School in East Greenwich, Rhode Island. He likes to have students collect real data, especially when the data can show connections between algebra and geometry.

Editorial Comment: See page 79 for an extension, "More Functions of a Toy Balloon."

to be straight lines, whereas others are curved. More advanced students can explore more precisely the mathematical models for each graph.

Too often, teachers are tempted to give students all the numbers they need to do a problem. The act of measuring is eliminated as an authentic part of the problem-solving experience. Students—and teachers—begin to take measurement for granted and assume that the procedure is easy. It usually is not. This activity demonstrates how each measuring tool presents its own challenges.

The data on diameter versus circumference, for example, *should* produce a nice example of direct variation, with the slope being a good approximation of π. Students familiar with the concept of π—that is, π is the ratio of circumference to diameter—will be surprised that the slope of the regression line is not 3.14 exactly. Does this discrepancy mean that the students have made a mistake? Probably not, and all the reasons for error—misreading the tape or the calipers, approximation error, recording error, irregular balloon shape, and so on—make fruitful topics for discussion, analysis, and assessment.

Some of the other relations—breaths versus diameter, for example—offer an opportunity to discuss the relationship between volume and radius. After graphs are drawn for these relations, a class discussion might flow along the following lines:

- Each breath represents a unit of volume.

- For a sphere,

$$\text{volume V} = \frac{4}{3}\pi r^3$$
$$= \frac{4}{3}\pi\left(\frac{d}{2}\right)^3$$
$$= \frac{\pi d^3}{6}.$$

- Solve for d in terms of V:

$$d = \sqrt[3]{\frac{6V}{\pi}}$$
$$= \sqrt[3]{\frac{6}{\pi}}\sqrt[3]{V}$$

- The diameter varies with the cube root of the volume.

- Since volume varies directly with the number of breaths, the fact that the diameter varies with the cube root of the number of breaths makes sense.

Each step in the discussion will involve quality class time in connecting algebraic and geometric concepts.

When students collect real data, they might not get what they expect. Remind them of the purpose of an experiment: why conduct an experiment at all if the results are already known? On the one hand, relations that should be linear may not look that way when graphed. On the other hand, the graph of breaths versus diameter may look fairly linear when it would be expected to be a cubic relation. These discrepancies do not mean that the experiment failed. To account for the unexpected results, students should discuss two key issues:

- What factors can cause variability in the data? Possible reasons include measurement error, variability in breath size, defects in balloon material, irregular stretching, fear of bursting the balloon with a normal breath, inconsistency in deciding the end of a flight, and so on.

- How could the experiment be changed to minimize these kinds of errors? Possible answers include the following: Use a commercial helium-balloon inflator to control each "breath" carefully. Use higher-quality balloons. Fly the balloons outside so they will not hit walls or furniture. Use two or more timers for each flight and average the results. Mark a spot on the balloon so that the diameter is consistently measured at the same place.

Sheet 3

Students should draw some of the graphs by hand because this process creates a better feel for the data. Technology, however, extends the analysis to an interesting new level of curve fitting and regression.

The table of values from this experiment offers ample opportunity to apply technology, especially curve-fitting programs in which one column of the table can be graphed against any other column. The TI-82 graphing calculator and the CA-Cricket Graph III (1993), for example, have this capability.

Furthermore, the discussion of curve fitting can lead to an interesting discussion of the measure of correlation between the data and the graph. We will call this correlation R. If the values of R and R^2 are available, these values should be interpreted intelligently, even though the actual derivation of the values need not be a part of the class discussion. Students, at some appropriate time, should learn that as $|R|$ approaches 1, the fit gets better; as R approaches 0, the function does a poorer job of describing the relationship. An R^2 value of 0.9 means that 90 percent of the data are accounted for by the function; a value of 0.7 means that 70 percent are accounted for, and so on (Freund and Smith 1986). It may be unsettling to discuss R without discussing exactly how it is calculated. In a sense, however, interpreting the value without fully understanding its source is like using a calculator without technically understanding its circuitry. Everyone does it all the time.

Students may discover that high-degree polynomials can thread through every point of a small data set—the higher the degree, the more bumps possible. Abusing polynomials is easy, which is an important lesson for students to learn in the age of accessible curve fitting. In discussing this hazard, teachers should point to the real connection among the data, the geometry, and the algebraic model. For the model of breaths versus flight time to be linear makes sense, since each breath should push the balloon for the same amount of time. It should make sense that breaths and diameter are related by the cube root of the radius by the nature of a sphere's volume. These same considerations argue against considering a logarithmic fit even when one gives a value of $|R|$ close to 1. The model should have real geometric and algebraic meaning.

Assessment

The activity is the assessment. How well did students manage the experiment? How good was the teamwork? Do the graphs match the data? Do the graphs show careful attention to detail? Do the responses reflect thoughtfulness and understanding?

An example of a chart with the data provided by a student group composed of Marianne San Antonio, Jenah Clabots, Rebecca Schwartz, Rebecca Quaglieri, and Michael Sullivan follows.

Number of Breaths Used to Inflate Balloon	Diameter (in cm)	Circumference (in cm)	Flight Time after Release
1	17.5	53	1.16
2	22	67	1.69
3	26	77	2.25
4	27	83	4.3
5	29	91	5.4
6	35	101	9.0

REFERENCES

Beckmann, Petr. *The Story of* π *(Pi)*. 3d ed. New York: St. Martin's Press, 1974.

CA-Cricket Graph III. Islandia, N.Y.: Computer Associates International, 1993. Software.

Freund, John E., and Richard Manning Smith. *Statistics: A First Course.* 4th ed. Englewood, N.J.: Prentice Hall, 1986.

Each group will need

- a **balloon inflator** first to practice taking even breaths and then to blow up the balloon;
- a **caliper user** to measure the diameter of the balloon;
- a **centimeter tape user** to measure the circumference of the balloon;
- a **timer** to measure the balloon's flight; and
- a **recorder** to gather the data in the chart below.

Instructions: Record the data generated by the activity. Inflate the balloon with one breath. Measure the diameter of the inflated balloon at its largest cross section, perpendicular to a line through the mouthpiece and the balloon's center, with the calipers. Use the centimeter tape to measure the distance between the ends of the calipers; measure the circumference of the cross section of the balloon. Release the balloon and time its flight until it "dies," not until it hits the ground. Repeat the activity but add one breath each time until the inflator fears that the balloon will burst.

No. of Breaths	Diameter	Circumference	Flight Time
1			
2			
3			
4			
5			
6			

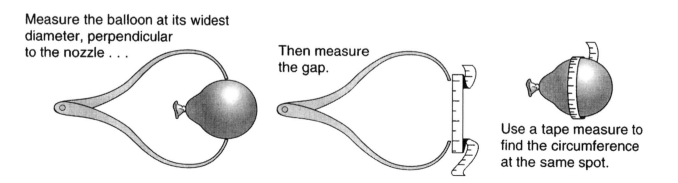

Measure the balloon at its widest diameter, perpendicular to the nozzle . . .

Then measure the gap.

Use a tape measure to find the circumference at the same spot.

From the *Mathematics Teacher*, November 1994

Sketch the following graphs from the data on sheet 1.

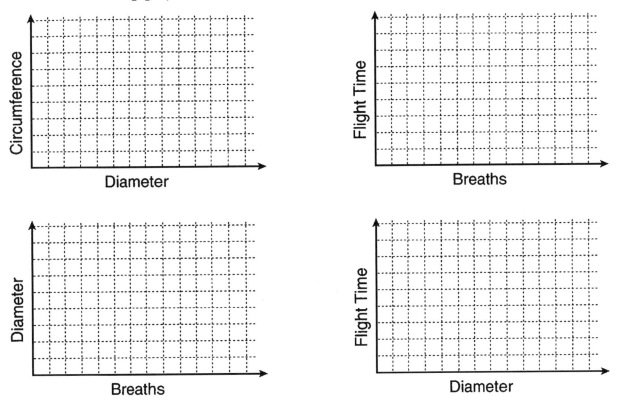

Discuss the graphs.

1. What do the graphs look like? Are they curved? Straight?_____

2. Which ones are similar? _____

3. Recall the concept of *slope*. Write down a definition._____

   _____

4. What does the slope of each graph mean here? _____

   _____

5. Do all the slopes stay the same? What does it mean when the slope changes within a
   graph? _____

   _____

6. How *precisely* were you able to make each measurement? _____

   _____

7. What would account for some of the measurement errors? _____

   _____

8. Write a function for each graph. _____

   _____

Use the data from sheet 1 and a graphing calculator or graphing software that handles tables of data to answer the following questions.

Diameter versus Circumference

1. Use the calculator or computer to do a linear regression on diameter (*x*) versus circumference (*y*). What does the *slope* of the regression line mean? What does the *intercept* represent? What does the value of *r* tell you? _____

2. What is the geometric definition of π? _____

3. Discuss the reasons for error. Why are the experimental values not exactly 3.14159 ...? What would make the experimental value too big? Too small? What is a reasonable level of precision given the tools and the real objects?

4. Discuss the value of π as (*a*) mentioned in the Bible (3), (*b*) established by the Egyptians,

$$\left(\frac{256}{81}\right),$$

and (*c*) calculated by Archimedes,

$$\left(\frac{22}{7} > \pi > \frac{223}{7}\right).$$

See Petr Beckmann's book *The History of π (Pi)* for a lively discussion of this number.

Breaths versus Diameter

5. Fit a power function to the graph of breaths (*x*) versus diameter (*y*). What do you find?

_____

6. Recall that the volume of a sphere is given with this formula:

$$V = \frac{3}{4}\pi r^3$$

(a) Solve this equation for *r*. _____

(b) Think about the power regression between breaths (*x*) and diameter (*y*). In the volume formula, which variable, *r* or *V*, corresponds most closely to the diameter? Explain. ___

_____

(c) In what sense is the number of breaths a measure of volume? Explain. _____

_____

7. Discuss ways to make these experiments more precise. What factors contribute to the validity of the data? _____

_____

Graphing Art

Fan Disher   February 1995

Edited by Timothy V. Craine, *Central Connecticut State University, New Britain, Connecticut;* Kim Girard, *Glasgow High School, Glasgow, Montana;* Guy R. Mauldin, *Science Hill High School, Johnson City, Tennessee*

TEACHER'S GUIDE

Introduction: Making connections between graphs and their equations to recognize equivalent representations of the same concept is essential for students to progress to higher mathematics (NCTM 1989, 146). In addition to the increased use of graphing calculators, students need experience analyzing and creating their own graphs and equations. Relevant practice improves their skill and self-confidence. The NCTM's *Curriculum and Evaluation Standards for School Mathematics* (1989) stresses the importance of students' developing personal self-confidence in their mathematical abilities and thereby gaining "mathematical power." To hold their interest, however, it is imperative to make this practice entertaining and challenging.

This activity was inspired by the many published activities in which students are given a set of equations and asked to graph them to form a picture. As valuable as these activities are for students, "[i]t is equally important, however, that they be given opportunities to translate from a graphical representation of a function to a symbolic form" (NCTM 1989, 155). In this activity, the student must draw a picture, analyze each line and curve, and write the symbolic equations represented by the graph. In addition to promoting analytical skills, the activity gives students an opportunity to express their creativity in a mathematical way.

Although this activity requires that the teacher spend some time assessing the students' work, the effort is worthwhile in that graphing skills are reinforced and students derive satisfaction from the experience. Students can work individually outside of class or in groups to create one picture per group.

Although this particular activity includes only lines, absolute-value graphs, and parabolas and was used in a second course in algebra, it can be extended to a more advanced project incorporating circles, ellipses, and hyperbolas. It can also be adapted for earlier work in algebra using only lines or for a more advanced course with sinusoidal curves.

Grade levels: 9–12

Materials: Sheet 1: rectangular-graph chalkboard or overhead projector-graph transparency and graph paper with x–axis –14 to 14 and y–axis –18 to 18. Sheet 2: sheet 4 and graphing calculator (optional). Sheet 3: sheet 4 and sheet 5, graph paper, and graphing calculator (optional).

Objective: After practicing graphing and equation writing, students will draw pictures on graph paper using only lines, absolute-value graphs, and parabolas. Students will then analyze each section of their graphs, write equations that will produce the graphs, and determine any restrictions on the domain and range for their equations.

Directions: Begin by reviewing the graphs of linear, quadratic, and absolute-value equations. Assign sheet 1, on which the students are given equations and asked to draw the graph. In addition to supplying graphing practice, this activity will introduce the concept of restricted domain.

On the second day, have students discover a general rule for graphing parabolas and absolute-value relations with a horizontal line of symmetry. Once students "relate procedures in one representation to procedures in an equivalent representation" (NCTM 1989, 146), they can usually derive their own rules for graphing forms of these equations even though the concept of inverse relation may not be introduced until later. Stress that these examples are no longer functions of x; therefore, they should restrict the y-values instead of the x-values. Assign sheet 2, a picture of a sailboat, for which they must determine the equations of each part of

Fan Disher currently teaches precalculus and AP calculus to gifted students at Mandeville High School, Mandeville, Louisiana, and cosponsors Mu Alpha Theta. In addition to promoting graphing-calculator use in her classroom, she experiments with other nontraditional activities to make mathematics enjoyable. She also presents mathematics and calculator workshops to mathematics teachers.

27

the picture and the restrictions on either the domain or the range. This group assignment promotes some very interesting discussions concerning the form of the equation and how to find the restrictions.

On the third day, have students compare their answers for the sailboats and discuss the fact that several correct forms for each equation are possible. Encourage them to be flexible by writing the equations in different forms as well as using statements with "and" and "or" in their equations. Also discuss different ways to write domain or range restrictions, including set notation, interval notation, and the method of combining restrictions with union (∪) notation. On sheet 2, they probably discovered that finding exact restrictions for their domains is not always easy. Review several ways to solve simultaneous equations or use graphing calculators to trace and find points of intersection. Practice both ways as necessary.

The students are ready to begin their projects using sheets 3 and 4. They should also receive the evaluation form on sheet 5 to let them know exactly how the project will be assessed. Some teachers may prefer to create a more holistic rubric to assess creativity as well as complexity and accuracy. If desired, place a limit on the number of equations students are allowed to use to make the process of assessment less timeñconsuming.

A rough draft of the picture with each line or curve numbered and at least ten of the equations should be due in about one week's time. At that time students can work in pairs to graph each other's equations and resolve some of the problems they may have encountered. Two weeks later, students should hand in final copies of both pictures and equations, their numbered rough drafts, and their self-evaluation on sheet 5 or another version. An easy way to assess the project is to give the equations to another student to graph and to discover the picture.

A FINAL PRODUCT

A final product can be a booklet containing all the graphs and equations. This compilation gives students a feeling of pride in their creation. (See students' work in figs. 1 and 2.) After all the equations have been checked for accuracy, appoint an editor from the class to oversee preparing the booklet for publication. Each student should correct any wrong equations and restrictions, copy equations neatly in ink on fresh copy of sheet 4, outline the picture with a fine marker, and name the picture. The class may vote to select a picture to be used as the cover of the booklet. Booklets can be photocopied and bound with plastic rings. The published booklet can be distributed to other mathematics teachers at the school to be used in their classes as models of excellent student work. Students will enjoy seeing their names and creations in print. They may not realize how much they have learned while they were having so much fun.

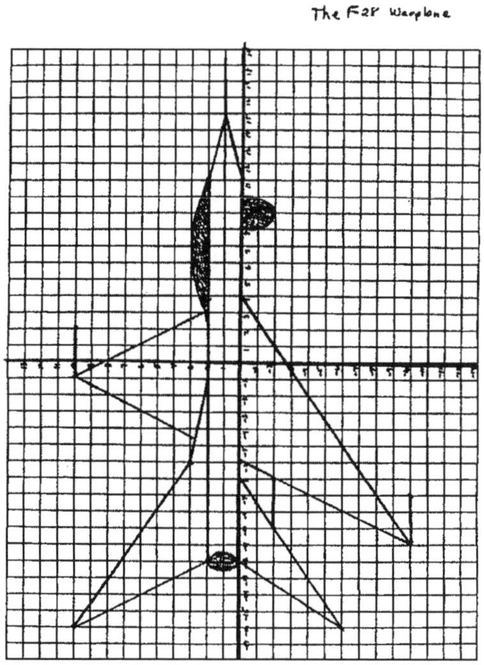

An F-28 war plane

Fig. 1. Student's equations, graph, and resulting creation

Using Activities from the *Mathematics Teacher* to Support *Principles and Standards*

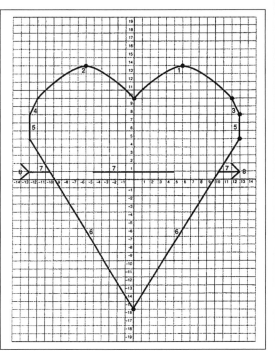

MATHEMATICIAN/ARTIST: _Tiffany_

EQUATIONS	RESTRICTIONS ON DOMAIN/RANGE
1. $y = -\frac{20}{121}(x - \frac{11}{2})^2 + 2$ or $y + 8 = \frac{20}{121}(x - \frac{11}{2})^2$	D [0, 11]
2. $y = 3$ or $y = 1$	D [$-\frac{11}{2}, -\frac{4}{2}$]
3. $x = (y+5)^2 - 11$	R [$-6, -4$]
4. $x = -(y+5)^2 - 9$	R [$-6, -4$]
5. $y = -(x+4)^2 - 4$ or $y + 6 = (x+4)^2$	D [$-5, -3$]
6. $x = -\frac{11}{2}$ or $x = -\frac{17}{2}$ or $x = -8$ or $y = -7$	R [1, 8]
7. $y = 1$ or $x = -2$ or $y = 5$	R [5, $\frac{13}{4}$]
8. $x = -8$ or $y = -11$ or $x = -14$	R [5, $\frac{7}{2}$]
9. $y = \frac{9}{12}$	D [$-14, 11$]
10. $Y = 11$	R [$-10, 5$]
11. $Y = 12$	R [0, 3]
12. $y = 3$ or $y = 0$	D [11, 12]
13. $y = -10$ or $y = 5$	D [$-14, 11$]
14. $x = -14$	R [$-10, 5$]
15. $y = -7$ or $y = -8$	D [$-13, -1$]
16. $Y = -13$ or $Y = -1$	R [$-8, -2$]
17. $y = 10a$	D [9, 10]
18. $y = 9$ or $Y = 10$	R [5, 16]
19. $y = -2x - 13$	D [$-8, -7$]
20. $y + \frac{15}{2} = 2|y+5|$ or $x = 2|y+5| - \frac{17}{2}$	R [$-6, -4$]
21. $x + \frac{13}{2} = 2|y+5|$	R [$-6, -4$]
22. $y = 2|x+6| + 2$	D [$-\frac{13}{2}, -\frac{11}{2}$]
23. $y = 3$ or $y = 2$ or $y = 1$	D [$-11, -10$]
24. $y = -11$	R [2, 3]
25. $x = -10$ or $x = -6$	R [1, 2]
26. $\{(-10, -5), (-5, -4)\}$	—

Sony radio

Fig. 2. Student's equations, graph, and resulting creation

Answers: Sheet 1: The graph produced from the given equations is shown in figure 3.

Fig. 3. Graphed solution to sheet 1

Sheet 2: The equations in table 1 produced the sailboat.

TABLE 1	
Equations	Restrictions on Domains and Ranges
1. $x = -2$	R: $[-8, -1] \cup [13, 16]$
2. $y = 15$ or $y = 16$	D: $[-2, 0]$
3. $x = 0$	D: $[15, 16]$
4. $y = -\frac{7}{5}|x + 2| + 13$	D: $[-12, 8]$
5. $x = 2(y - 10)^2 - 3$	R: $[9, 11]$
6. $y = -\frac{1}{3}x + 7$	D: $[5.08, 3]$
7. $y - 6 = -\frac{1}{3}(x + 3)$	D: $[-6.23, 4.88]$
8. $x + 3y = 10$	D: $[-7.19, 6.44]$
9. $y = \frac{1}{18}(x + 2)^2 - 16$	D: $[-14, 10]$
10. $y = -1$	D: $[-12, 8]$
11. $y = -8$	D: $[-14, 10]$

REFERENCE

National Council of Teachers of Mathematics. *Curriculum and Evaluation Standards for School Mathematics.* Reston, Va.: The Council, 1989.

Graphing Art

Graph 1

Directions: Graph the following equations on one set of coordinate axes. Make sure to stay within the indicated domains (D) and ranges (R).

Equations	Restrictions on Domains and Ranges
1. $y = -\dfrac{1}{9}(x-6)^2 + 14$	D: [0, 12]
2. $y = -\dfrac{1}{9}x^2 - \dfrac{4}{3}x + 10$	D: [-12, 0]
3. $y = -2x + 34$	D: [12, 13]
4. $2x - y = -34$	D: [−13, −12]
5. $x = 13$ or $x = -13$	R:[5, 8]
6. $y = \left\| \left(\dfrac{21}{13}\right)x \right\| - 16$	D: [−13, 13]
7. $y = 1$	D: [−13, −10.5] $\cup$ [−5, 5] $\cup$ [10.5, 13]
8. $x = -\|y - 1\| + 13$ or $x = -\|y - 1\| - 13$	R: [0, 2]

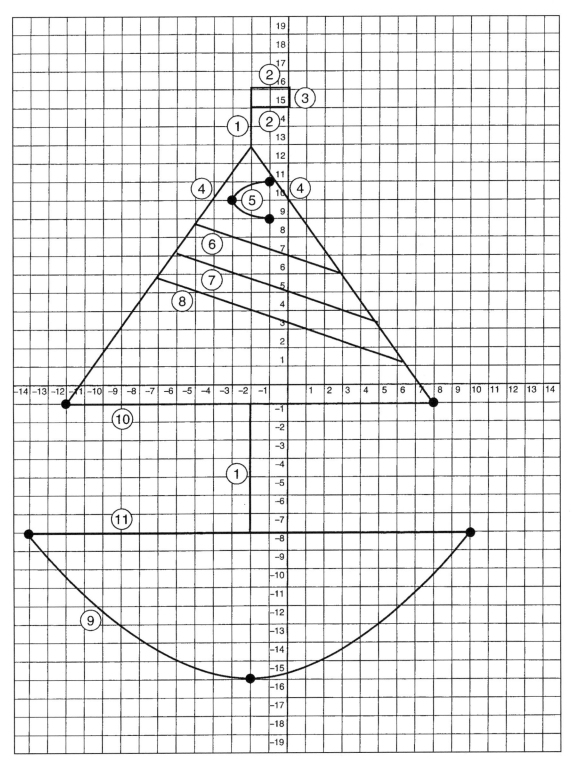

Directions

1. Find at least one equation for each section of the graph from 1 through 11.

2. Use three different *forms* of equations for sections 6, 7, and 8.

3. Use domain or range restrictions and compound statements using "and" and "or" where appropriate.

From the *Mathematics Teacher*, February 1995

Directions

1. Using graph paper, draw a picture containing only graphs of lines, parabolas, and absolute values. The picture must contain graphs of a minimum of ten (10) different equations. At least two must be lines; at least two, parabolas; and at least two, absolute-value graphs. One of the absolute-value graphs or one of the parabolas must have a horizontal line of symmetry. The remaining equations may be any of the three.

2. Domains and ranges may be determined by solving equations simultaneously or by using the graphing calculator to graph the functions and trace the curves to find points of intersection. If these points are not integers, specify them to the nearest hundredth.

3. Make two copies of your graph. Hand in a rough copy of your final figure with the different parts of your graph numbered to coincide with the equations. Later you will hand in an unnumbered final copy of your picture that has been outlined in a black fine-tip marker. Hand in the equations printed neatly in black ink on the correct form. Name your picture and write the name at the top of your equation page.

Evaluation

Attached is the sheet that both you and I will use to evaluate your project. Fill in this sheet and hand it in with your final copy.

Timetable

1. A rough-draft picture with at least ten of the equations is due on _____ .

2. The final pictures, all equations in proper form, and the completed evaluation sheet are due on_____.

CREATING A GRAPH

SHEET 4

Title ______________________________

Mathematician/Artist ______________________________

Equations	Restrictions on Domain and Range
1.	
2.	
3.	
4.	
5.	
6.	
7.	
8.	
9.	
10.	
11.	
12.	
13.	
14.	
15.	
16.	
17.	
18.	
19.	
20.	
21.	
22.	
23.	
24.	
25.	

From the *Mathematics Teacher*, February 1995

Name_____

Date _____ Hour_____

Graphing Art I

Lines, Parabolas, and Absolute Values

Equations of parabolas

_____ no. correct/_____ total no. of parabolas = _____ × 25 = _____

Equations of absolute-value graphs

_____ no. correct/_____ total no. of absolute-value graphs = _____ × 25 = _____

Equations of lines

_____ no. correct/_____ total no. of lines = _____ × 25 = _____

Restrictions on domains/ranges

_____ no. correct/_____ total no. of restrictions = _____ × 25 = _____

Complexity

Rubric for Assigning Points for Complexity:

A maximum of 10 points can be earned in any of the following ways:

	0 points	1 point	2 points	3 points
No. of equations used	10	11–15	16 – 25	over 25
No. of complex domains/ranges	0	1	2	over 2
No. of equations in different forms	0	1– 2	3 – 4	over 4
No. of horizontal graphs (one is required)	1	2	3	over 3
No. of uses of union in D/R or compound-sentence equations	0	1– 4	5 – 10	over 10

Rating: 1–10 = _____

Form (neatness, proper labeling, numbered rough draft, etc.)

Teacher comments on form:

_____

_____

_____

Rating: 1–5 = _____

Total: = _____

Time for Trigonometry

Harris S. Shultz and Martin V. Bonsangue   May 1995

Edited by Timothy V. Craine, *Central Connecticut State University, New Britain, Connecticut;* Kim Girard, *Glasgow High School, Glasgow, Montana;* Guy R. Mauldin, *Science Hill High School, Johnson City, Tennessee*

TEACHER'S GUIDE

Introduction: Understanding and describing periodic motion are important topics for an increasing number of disciplines in the natural and social sciences. The *Curriculum and Evaluation Standards for School Mathematics* (NCTM 1989, 163) states that circular functions "are mathematical models for many periodic real-world phenomena, such as uniform circular motion, temperature changes, bio-rhythms, sound waves, and tide variations." Although periodic behavior occurs naturally in many situations, the mathematical study of periodic functions is often abstract and related to physical examples indirectly, if at all. This article describes an activity introducing periodic behavior in which students generate graphs of circular functions using a familiar physical model.

In Barrett and Bartkovich et al. (1992), periodic behavior is introduced by asking students to describe the shape of the graph of distance versus time where the dependent variable is the distance between the ceiling and the tip of the hour hand of a clock mounted on the wall. The graph turns out to be a transformation of the graph of $d = \sin t$ and can be described by an equation of the form

$$d = b + A \sin (30 (t - c)),$$

where the constants A, b, and c depend on the radius of the clock, the distance from the ceiling to the center of the clock, and the position of the hour hand at the initial time, respectively. In this representation, the argument of the sine function is expressed in degrees.

This introduction to the sine function, and, of course, to the cosine function, gives students a "hands on" basis for these functions, certainly more than the abstract notion of a point moving about a theoretical unit circle. The approach is impressive and can be extended to an enriching activity.

Grade levels: 9–12

Materials: Activity sheets, scissors, centimeter rulers, scientific calculators, graphing calculators (optional)

Directions: Sheets *1A* and *1B*: For clocks 1, 2, and 3, students measure the distance from the ceiling, represented by the horizontal line above the clock, to the tip of the hour hand, assumed to be on the circle itself, at each hour beginning at 12 o'clock, which is taken to be time $t = 0$. For each time t, they record the results of their measurements in the column under d. For clock, 4, $t = 0$ is taken to be 3 o'clock. Thus, $t = 1$ corresponds to 4 o'clock, and so on.

Sheet 2: For each of the four clocks on sheets 1A and 1B, the ordered pairs (t, d) are used to create a scatterplot. Students should recognize the periodicity of the function d as they plot values for $t = 13, 14, \ldots , 24$. When the points of each scatterplot are smoothly connected, initial glimpses of a sine curve are obtained. Students should recognize that the amplitude of the graph for clock 2 is less than the amplitude of the graph for clock 1. They should recognize that the graph for clock 3 is a translation 1.5 units upward of the graph for clock 1. They also should recognize that the graph for clock 4 is a translation three units to the left of the graph for clock 1.

Sheet 3: At this point students are prepared to predict the shapes of the graphs obtained when various-sized clocks are placed at various locations on the wall and when various positions of the hour hand are taken as the initial starting point. The graphs for clocks 5, 6, 7, and 8 should be sketched without making any measurements. That is, students should be encouraged to predict the shapes of the graphs on the basis of patterns observed in completing sheet 2.

Harris Shultz teaches at California State University at Fullerton, Fullerton, California. In 1992, he received the Distinguished Teaching Award from the Southern California Section of the Mathematical Association of America. Marty Bonsangue teaches at California State University at Sonoma, Rohnert Park, California. He enjoys helping elementary and secondary school teachers find mischievous ways to engage their students in mathematics.

Sheet 4: Students are now ready to explore on their own the formal development of the sine function by using a clock having radius equal to one unit. To fill in the column headed *X*, they must use the fact that in one hour, the hour hand of a clock generates an angle measuring 30 degrees. Once the second column of the chart is completed, students should cut out the ruler, verify that the clock indeed has a radius of one unit; and then complete the third column, measuring the distance *d* from the line segment joining the 12 and the 6 on the clock to the tip of the hour hand. With more advanced students, radian measure can be used in the second column.

Sheet 5: Next, the sine function is formally introduced. A scientific calculator, set to degree mode, should be used by students to check the measurements recorded in the *d* column on sheet 4. These measurements should closely agree with the values sin 30k, k = 0, 1, 2, . . . , 12. A graphing calculator set to degree mode and having a table option, or almost any spreadsheet program, can verify that the formulas $d = 3.5 + 3 \sin (30 (t - 3))$, $d = 3.5 + 2 \sin (30 (t - 3))$, $d = 5 + 3 \sin (30 (t - 3))$, and $d = 3.5 + 3 \sin (30t)$ yield ordered pairs very close in value to the data collected for clocks 1 through 4, respectively.

Ideas for assessment: These activities furnish a natural setting to incorporate alternative assessments of the students' understanding of circular functions (see NCTM [1991]). Such assessment could include the following components:

- Have students include sheet 2 in their portfolios, along with comparable problems that they have created and solved themselves. Students should write or use pictures to explain their solutions. Have students trade papers and identify the starting time of the clock from graphs similar to those in sheet 3.

- Introduce the words *amplitude, period,* and *phase shift* by having students describe these characteristics in their own words.

- In whole-class discussion, ask students to identify real-world examples of circular phenomena, such as the distance above the ground on a Ferris wheel or the reflector on the rim of a bicycle wheel. Ask students to identify what the circular function would measure in each example and why this information is important.

- Working together in teams of three or four, students can be assigned a week-long project creating a mathematical model of circular phenomena identified in the discussion. For some models, the circular function would be generated with the distance of the moving point above the ground rather than below a "ceiling."

- Working together in teams of two, students can use their graphing calculators to explore changes in the graphs of the equation

$$d = b + A \sin (k(t - c))$$

by varying values for the parameters *b, A, c,* and *k*. This exploration not only involves the students in thinking about how the graph is affected by changes in the parameters but gives the teacher insight into the students' logic in testing each parameter. Do they change only one parameter while holding the others constant, or do they try to make conclusions on the basis of simultaneous "experimental" changes? The students' use or nonuse of controls could serve as a springboard to a discussion of effective and ineffective approaches to mathematical modeling.

- Ask students to write about how and why the graphs are changed as a result of varying each parameter. Look for the students' understanding of variations in amplitude, period, and phase as independent changes.

- For further exploration, have students consider the *horizontal* distance from the hour hand to a vertical line, such as a wall. Use this example to discuss the notion of a function that is correlated to that function describing the vertical distance. This activity furnishes background information for both the nature and name of the cosine function.

Answers: Sheets 1A and 1B:

Clock 1		Clock 2		Clock 3		Clock 4	
t	d	t	d	t	d	t	d
0	0.5	0	1.5	0	2.0	0	3.5
1	0.9	1	1.8	1	2.4	1	5.0
2	2.0	2	2.5	2	3.5	2	6.1
3	3.5	3	3.5	3	5.0	3	6.5
4	5.0	4	4.5	4	6.5	4	6.1
5	6.1	5	5.2	5	7.6	5	5.0
6	6.5	6	5.5	6	8.0	6	3.5
7	6.1	7	5.2	7	7.6	7	2.0
8	5.0	8	4.5	8	6.5	8	0.9
9	3.5	9	3.5	9	5.0	9	0.5
10	2.0	10	2.5	10	3.5	10	0.9
11	0.9	11	1.8	11	2.4	11	2.0
12	0.5	12	1.5	12	2.0	12	3.5

Sheet 2:

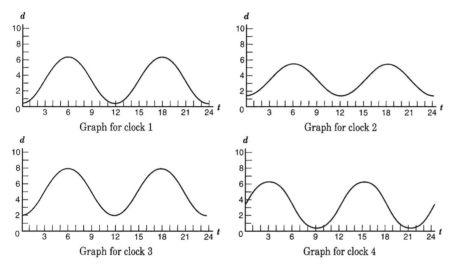

Graph for clock 1

Graph for clock 2

Graph for clock 3

Graph for clock 4

Sheet 3:

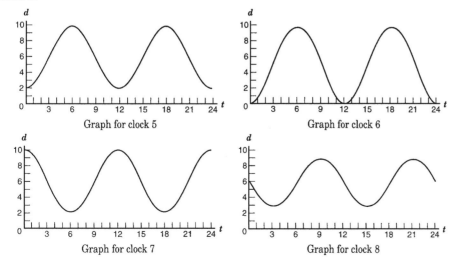

Graph for clock 5

Graph for clock 6

Graph for clock 7

Graph for clock 8

t	X	d
0	0°	0.0
1	30°	0.5
2	60°	0.9
3	90°	1.0
4	120°	0.9
5	150°	0.5
6	180°	0.0
7	210°	−0.5
8	240°	−0.9
9	270°	−1.0
10	300°	−0.9
11	330°	−0.5
12	360°	0.0

Sheet 5: The formula $Y_1 = 5 + 3 \sin (30 (X - 3))$ will yield the table obtained for clock 3, and the formula $Y_1 = 3.5 + 3 \sin (30X)$ will yield the table obtained for clock 4.

REFERENCES

Barrett, Gloria B., Kevin G. Bartkovich, et al. *Contemporary. Precalculus through Applications.* Dedham, Mass.: Janson Publications, 1992.

National Council of Teachers of Mathematics. *Curriculum and Evaluation Standards for School Mathematics.* Reston, Va.: The Council, 1989.

——. *Professional Standards for Teaching Mathematics.* Reston, Va.: The Council, 1991.

Clock 1: The clock below has radius equal to 3 cm. Its center is 3.5 centimeters below the "ceiling." Measure the distance *d* from the ceiling to the tip of the hour hand at each hour, beginning at 12 o'clock, which is taken to be time *t* = 0. Record your results in the chart.

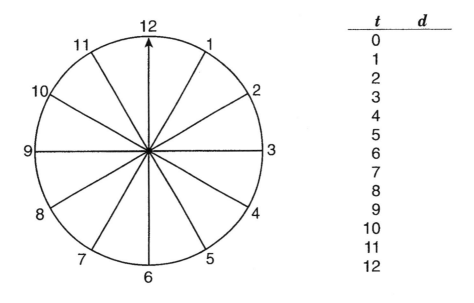

t	d
0	
1	
2	
3	
4	
5	
6	
7	
8	
9	
10	
11	
12	

Discussion question: What values of *d* would you get for *t* = 13, 14, . . . , 24?

Clock 2: The clock below has radius equal to 2 cm. Its center is 3.5 centimeters below the "ceiling." Measure and record the distances as you did for clock 1.

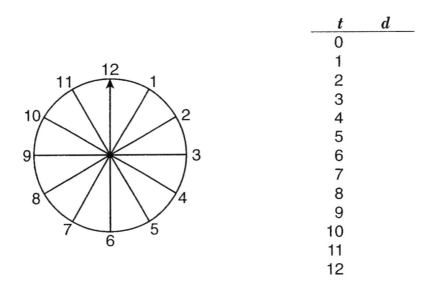

t	d
0	
1	
2	
3	
4	
5	
6	
7	
8	
9	
10	
11	
12	

Clock 3: The clock below has radius equal to 3 cm. Its center is 5 centimeters below the "ceiling." Measure and record the distances as you did for clocks 1 and 2.

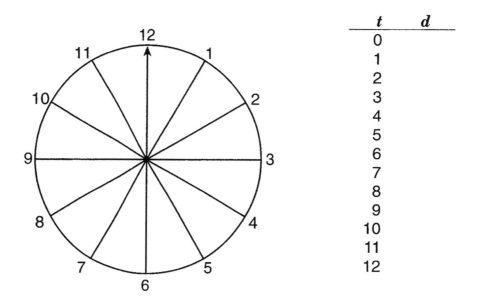

t	d
0	
1	
2	
3	
4	
5	
6	
7	
8	
9	
10	
11	
12	

Clock 4: The clock below has radius equal to 3 cm. Its center is 3.5 centimeters below the "ceiling." Measure and record the distances as you did for clocks 1, 2, and 3, but begin at 3 o'clock, which is taken to be time $t = 0$.

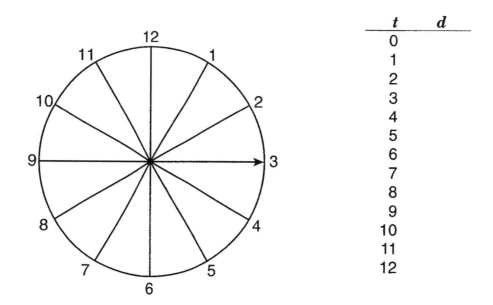

t	d
0	
1	
2	
3	
4	
5	
6	
7	
8	
9	
10	
11	
12	

From the *Mathematics Teacher*, 1995

For each of the four clocks on sheets 1A and 1B, create a scatterplot for time $t = 0$, 1, 2, 3, . . . , 24 using the recorded data. Then connect the points with a smooth graph.

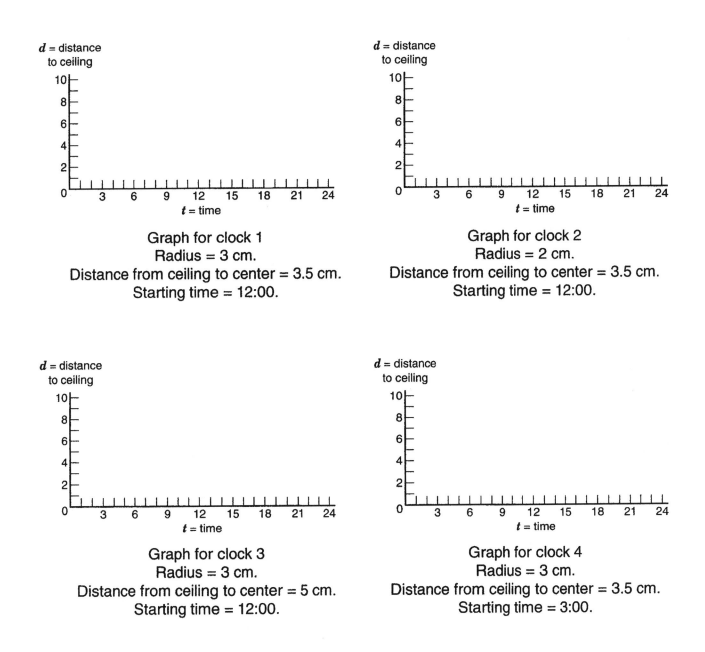

Graph for clock 1
Radius = 3 cm.
Distance from ceiling to center = 3.5 cm.
Starting time = 12:00.

Graph for clock 2
Radius = 2 cm.
Distance from ceiling to center = 3.5 cm.
Starting time = 12:00.

Graph for clock 3
Radius = 3 cm.
Distance from ceiling to center = 5 cm.
Starting time = 12:00.

Graph for clock 4
Radius = 3 cm.
Distance from ceiling to center = 3.5 cm.
Starting time = 3:00.

Discussion question: Discuss how the four graphs are alike and how they are different.

Sketch the graph of the distance of the hour hand from the ceiling as a function of time for each clock described below.

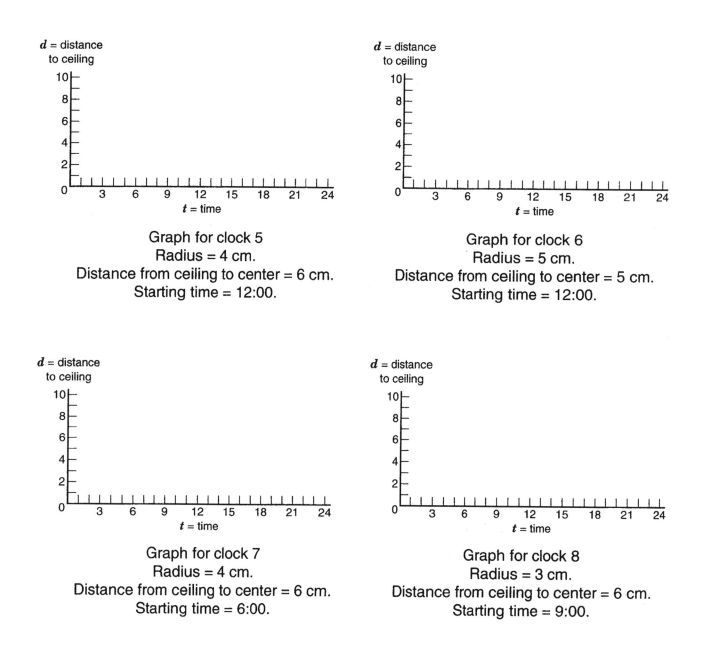

Graph for clock 5
Radius = 4 cm.
Distance from ceiling to center = 6 cm.
Starting time = 12:00.

Graph for clock 6
Radius = 5 cm.
Distance from ceiling to center = 5 cm.
Starting time = 12:00.

Graph for clock 7
Radius = 4 cm.
Distance from ceiling to center = 6 cm.
Starting time = 6:00.

Graph for clock 8
Radius = 3 cm.
Distance from ceiling to center = 6 cm.
Starting time = 9:00.

Discussion question: Explain how you made your predictions.

The clock shown here has a radius of one unit. Introduce a new variable, namely, the angle X, generated by the hour hand beginning at 12 o'clock, which is taken to be time $t = 0$. Complete the second column of the chart.

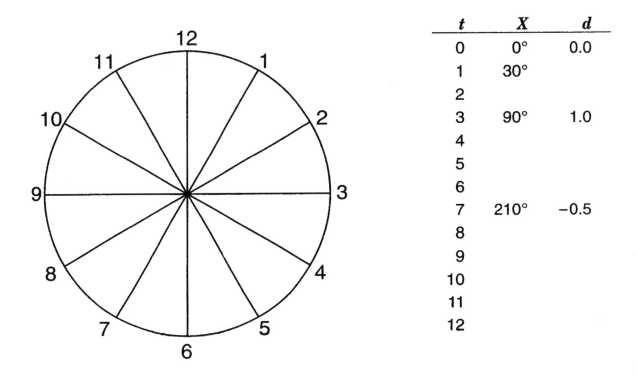

t	X	d
0	0°	0.0
1	30°	
2		
3	90°	1.0
4		
5		
6		
7	210°	−0.5
8		
9		
10		
11		
12		

Cut out the ruler at the bottom of the page. Use this unit ruler to check that the radius of the clock is actually one unit. Use the ruler to measure the distance d from the vertical line through the center of the circle to the tip of the hour hand. Let d be positive when the hour hand is to the right of the vertical line and negative when it is to the left. Begin measuring at 12 o'clock, which will be time $t = 0$. Record your results in the third column of the chart.

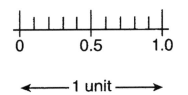

0   0.5   1.0

←——— 1 unit ———→

From the *Mathematics Teacher*, 1995

On sheet 4, the function that expresses d in terms of X is called the sine function. Thus, if X is the angle generated by the hour hand beginning at 12 o'clock, then the sine of X is the perpendicular distance from the tip to the vertical line through the center of the circle. Write this equation as $d = \sin X$.

1. Use a scientific calculator set to degree mode to compute $\sin 0°$, $\sin 30°$, $\sin 60°$, and so on. Compare each value with the value of d that you recorded in the chart on sheet 4. What do you find?

2. If you have a graphing calculator with table capabilities, enter $3.5 + 3 \sin (30 (X - 3))$ for Y_1. Begin the chart with $X = 0$ and use increments of 1. Compare the chart with the results that you obtained for clock 1 on sheet 1A, making sure that the calculator is set to degree mode. What do you find?

3. Repeat this procedure using $3.5 + 2 \sin (30 (X - 3))$ for Y_1, and compare the chart with the results that you obtained for clock 2 on sheet 1A. What do you find?

4. What formulas for Y_1 would yield the charts that you obtained for clocks 3 and 4 on sheet 1B?

What Is the *r* For?

Loring Coes III

December 1995

Edited by Timothy V. Craine, *Central Connecticut State University, New Britain, Connecticut;* Kim Girard, *Glasgow High School, Glasgow, Montana;* Guy R. Mauldin, *Science Hill School, Johnson City, Tennessee;* Henri Picciotto, *The Urban School, San Francisco, California*

TEACHER'S GUIDE

The following scene from an intermediate-algebra class points out the new power of technology in the modern classroom as well as questions that arise because of it.

Meliza, Graham, and Courtney used a computer with an overhead display to graph the following data (Brown 1992, 683) about world bicycle production over the last four decades.

Decade	*x*	*y*	World Bicycle Production in Millions
1960	1	20	
1970	2	36	
1980	3	62	
1990	4	95 (estimate)	

Next they did a linear regression to find a line that fit the data well. Using Cricket Graph III (Computer Associates International 1993), the computer showed the following graph.

Should we use a computer function if we do not know how it works?

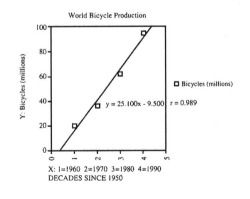

World Bicycle Production

$y = 25.100x - 9.500$ $r = 0.989$

X: 1=1960 2=1970 3=1980 4=1990
DECADES SINCE 1950

"The data look a little curvy, but the line fits the points pretty well," Becky said.

"The 25.10 is the slope and it means that production increased by about 25 million every decade," Marcus added.

"Yes, but the −9.5 says that in 1950 [the graph at 0], there were −9.5 million bikes made. That doesn't make any sense," Naomi observed.

"You're right, but the line does work pretty well from 1960 to 1990," Kate said.

"What's the *r* for?" Sarah asked.

"And what does *r* = 0.989 mean?" Jared wanted to know.

The students' questions about *r*, the correlation coefficient, pose a dilemma, unknown a generation ago, for mathematics teachers. Should we use a computer function if we do not know how it works? How much of the technology and underlying mathematics must students understand before they can use this built-in computing power?

Many teachers I know believe that using electronic tools is a mild, modern kind of cheating. On a more serious level, as Dubinsky (1995) points out, technology can be overused, and errors can occur if students do not have the mathematical understanding to support the technology. This issue is rich and interesting, and one that we all need to discuss.

It is important for us to resolve this issue for many reasons. The computing power at our fingertips is immense and can widen mathematical horizons in our classrooms. To shun the advanced functions is unrealistic; they are there, and we will use them. And finally, this dilemma will only recur as technology continues to improve.

Loring (Terry) Coes III is chair of the mathematics department at Rocky Hill School in East Greenwich, Rhode Island. He is interested in the impact of technology on education.

Editorial comment: Some calculators provide the value of r with the regression equation. Others store it as a statistics variable that may be retrieved. For example, after calculating a regression equation on a TI-83 Plus, use VARS, 5: Statistics, EQ, 7: r to find the value of r.

I believe, moreover, that reasonable and highly productive solutions to this dilemma are possible. We have handled equivalent problems for a long time, but in different settings. Most of us drive a car without fully understanding its inner workings. We have a general sense of how it is powered and how the steering system works, but we can drive without concentrating on the engineering details. We can be good drivers without being good engineers or good mechanics. We clearly understand that a driver requires a less sophisticated understanding of automobile mechanics than an engineer. This analogy is appropriate to some of the advanced mathematical functions students currently have at their fingertips.

Most of us drive a car without fully understanding its inner workings

The activities in this article are a practical response to the philosophical debate about the use of technology. The focus is on r and its interpretation for high school students. Students will learn how to interpret r and will also see that sophisticated mathematics embedded in r can be better understood with further study.

SOME BACKGROUND ON r

The variable r is also called the *correlation coefficient* and measures how well a line or curve fits a set of data points. Its value varies from –1 to +1. Positive values of r indicate that the line has a positive slope; negative values indicate a negative slope. When r is close to 0, little or no association, or no functional relationship, exists between x and y. When the absolute value of r is close to 1, a strong association is found between the two variables.

Several formulas are possible for computing r. *Data Analysis and Statistics across the Curriculum* (Burrill et al. 1992) gives one of them,

$$r = \frac{\sum_{i=1}^{n}(\overline{x}-x_i)(\overline{y}-y_i)}{(n-1)S_xS_y},$$

where S_x and S_y are the standard deviations of x and y, respectively. The standard deviation is a measure of the dispersal of the data around the mean.

Another way to think of the correlation coefficient is that the value of r is a function of the "error," that is, the vertical distances between the points and the fitted line. See figure 1.

An r-value close to ±1 shows that the line describes the data well. As the students saw with the bicycle-production data, however, even a good r-value does not ensure that the line describes the phenomenon accurately outside the range of data.

The bicycle line may not be a good predictor of the future or representation of the past.

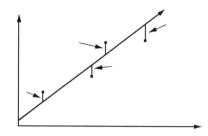

Fig. 1. The value of r is a function of these distances from the best-fitting line.

Most books on statistics offer a more detailed discussion of r. See *Statistics: A First Course* (Freund and Smith 1986) for one example.

Grade levels: 9–12

Materials: Graph paper; a graphing calculator, such as a TI-82, or a graphing program, such as Cricket Graph III

Procedure: Have students work in small groups and compare their responses. Many items are open ended and can have various solutions.

Sheet 1: Students will get some paper-and-pencil experience with creating linear regressions and will discuss the criteria for a good linear fit. Teachers may want to suggest at an appropriate point in the discussion that minimizing errors or the distances from the line is important. Linear regression is actually based on minimizing the sum of the *squares* of the errors.

To get r close to 0, try placing four points as the vertices of a square and the fifth point in the center of the square. Students will also need to learn how to do a linear regression with their existing technology.

Sheet 2: Students will have to experiment with different points to find data sets that fall within the requested ranges. It will be especially interesting to see how many different ways students can create r-values within the given ranges.

Students can find a discussion of r in different textbooks. The formula for r is not important yet. At this stage it is most important for students to learn that r varies from –1 to +1 and to *interpret* the value of r as a measure of how well the data fit the regression line.

Sheet 3: Students should discuss different ways of making r equal to 0. Teachers may want to emphasize that in doing a linear regression, we are trying to get a function to conform to the relationship $y = ax + b$, where y is a linear function of x. As r gets close to 0, the functional relationship weakens.

As students complete problems 4 and 5, where points are outside the range $0 \leq x \leq 10$ and $0 \leq y \leq 10$, it becomes clear that the *relative* distances between the points are very important.

Sheet 4: Students are asked to think about the philosophical implications of using a computing tool without fully understanding it. Students should come to realize that we can, to a surprising degree, handle machines and tools even though we do not understand all about their inner workings.

A good extension to the questions contained on this sheet is to assign chapter 10 of *Gödel, Escher, Bach: An Eternal Golden Braid* (Hofstadter 1980). Students can then discuss how Hofstadter's ideas about the relationship between people and computers are related to using the r-values before we know how r is derived.

Assessment: The activities are the assessment. Teachers should remember that the focus of these activities is the *interpretation* of r-values, not their calculation.

Students should be able to visualize a data set that might have an r-value close to 1, close to -1, and close to 0. Students should understand that a single r-value can represent several different data sets. To create two different sets that have r equal to 1, for example, create two different collinear sets with different slopes.

Answers to selected problems: Sheet 1: 2. An equation like the following will fit the points fairly well: $y = 0.8x + 1.5$. 4. Try moving the two "low" points higher: change (2, 2) to (2, 4) and (7, 5) to (7, 8). 7. $y = 0.534x + 4.0465$; $r = 0.6628$. 8. One possibility is to change (3, 7) to (3, 5). 9. One possibility is to change (6, 6) to (6, 7). 10. As the points cling more closely to the regression line, r becomes closer to 1.

Sheet 2: 1.–5. Answers will vary. Here is one possible set.

1. (1, 5), (2, 7), (3, 6), (4, 7), (3, 10)
2. (2, 1), (2, 4), (5, 2), (5, 4), (4, 7)
3. (2, 10), (2, 6), (5, 2), (5, 4), (4, 7)
4. (2, 3), (2, 6), (5, 2), (5, 4), (4, 6)
5. (2, 1), (2, 8), (5, 1), (5, 8), (3, 6)

In assessing this sheet, teachers may want to inspect the students' calculators directly, which will be easier than testing each answer set separately.

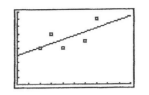

6. r gets closer to +1 as the data better fit a line with positive slope.
7. r gets closer to 0 as the data points become more random.

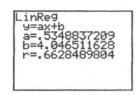

8. r gets closer to -1 as the data better fit a line with negative slope.

Sheet 3:
3. $y = 0.686x + 2.86$; $r = 0.3663$.

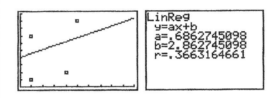

4. One possibility is (100, 72).
5. Given the answer to (4), one possibility is (−200, 200).

Sheet 4: 3. Possible responses may include finding the point of intersection of two curves using the "intersect" command, finding the result of raising a positive number to a nonintegral power, and computing values of the trigonometric functions. 4.-6. Example: We can program a VCR to tape a show next Tuesday at 9:00 P.M. as long as we program the correct sequence of instructions for the VCR. We do not need to know anything about the circuitry, or *internal* programming, of the VCR system. We do not need to know about anything "inside the box."

Students can use r-values before knowing how r is derived

BIBLIOGRAPHY

Brown, Lester R., Christopher Flavin, and Hal Kane. *Vital Signs 1992*. New York: World Watch Institute and W. W. Norton & Co., 1992.

Brown, Richard G. *Advanced Mathematics*. Boston: Houghton Mifflin Co., 1992.

Burrill, Gail, John C. Burrill, Pamela Coffield, Gretchen Davis, Jan de Lange, Diann Resnick, and Murray Siegel. *Data Analysis and Statistics across the*

Curriculum. Addenda Series, Grades 9–12. Reston, Va.: National Council of Teachers of Mathematics, 1992.

Computer Associates International. Cricket Graph III. Islandia, N.Y.: Computer Associates International, 1993. Software.

Dubinsky, Ed. "Technology Tips: Is Calculus Obsolete?" *Mathematics Teacher* 88 (February 1995): 146–48.

Freund, John E., and Richard Manning Smith. *Statistics: A First Course*. Englewood Cliffs, N.J.: Prentice Hall, 1986.

Hofstadter, Douglas R. Gödel, Escher, Bach: *An Eternal Golden Braid*. New York: Vintage Books, 1980.

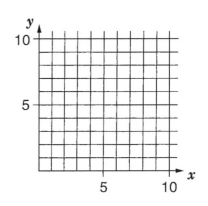

1. Plot the following points on the grid: (2, 2), (5, 7), (7, 5), (8, 9), (1, 3).

2. Look at the graph and discuss how to draw a straight line that fits this set of points.

3. When you are trying to draw a line through a set of points, what makes a good fit? What makes a poor fit?

4. Suppose that you could move two points so that the set would more closely fit a line. Which two points would you move and how? Plot the points on graph (a).

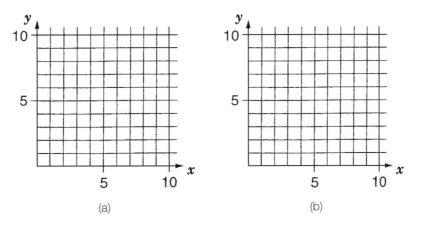

(a)   (b)

5. On graph (b), show how you can spread five points around so that no line will fit them well. How did you do it?

6. Use appropriate technology to plot the following points: (4, 5), (3, 7), (2, 5), (7, 9), (6, 6). Sketch the points on the grid.

7. Use the calculator or computer to do a linear regression on the data points. Record the equation and the *r*-value that you get. Graph the linear regression line along with the points.

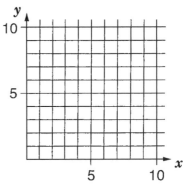

   Equation:_____ *r* = _____

8. Move one point in the data set in question 6 so that *r* > 0.8. Circle the point you moved and mark its new location on the graph with an *X*.

9. Move an additional point so that *r* > 0.9. Circle the point you moved and mark its new location on the graph with an *X*?

10. What appears to make *r* get closer to +1?

In the following problems, choose five points in the window $0 \leq x \leq 10$, $0 \leq y \leq 10$ so that r falls in the given ranges:

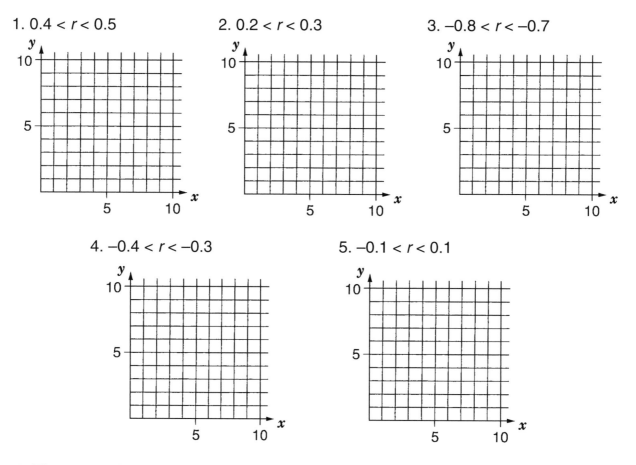

1. $0.4 < r < 0.5$

2. $0.2 < r < 0.3$

3. $-0.8 < r < -0.7$

4. $-0.4 < r < -0.3$

5. $-0.1 < r < 0.1$

6. The value of r varies from -1 to $+1$. What causes r to get close to $+1$?

7. What causes r to get close to 0?

8. What causes r to get close to -1?

9. The value of r is called the *correlation coefficient*. Discuss what r tells you about the points and about the regression line.

10. Look up "correlation coefficient" in a book, such as *Advanced Mathematics* (Brown 1992). Do your answers for questions 6–9 agree with what you found?

1. Select five different points in two different ways so that $r = 0$.

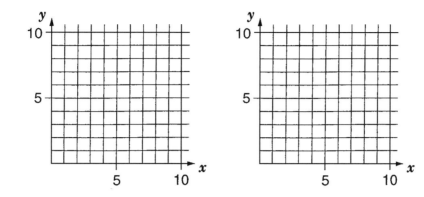

2. What do your two selections have in common?

3. Plot the points (1, 1), (1, 7), (4, 2), and (5, 9). Do a linear regression and find the r-value.

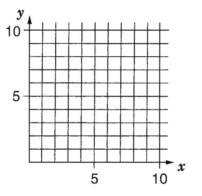

4. Add a fifth point, whose x- and y-coordinates can be greater than 10 or less than 0, so that the new r-value is at least 0.99.

5. Add a sixth point so that $r < -0.7$.

6. Explain why adding points outside the $0 \le x \le 10$, $0 \le y \le 10$ window can have such an impact on the value of r.

1. Look up a *formula* for *r* in the school library or in a textbook. Write it here.

   _____

2. In what course might you study correlation in more detail?

   _____

3. Describe two other functions on a calculator that you use without fully understanding the mathematics behind the calculator's algorithm for arriving at its answer.

   _____

   _____

   _____

4. Describe two situations in which you know how to use a machine without fully understanding how it works.

   _____

   _____

   _____

5. Think about your answers to problem 4. For each situation, describe what you need to know to use the machine to its fullest.

   _____

   _____

   _____

6. For each situation in problem 4, describe what you *do not* need to know.

   _____

   _____

   _____

The Inverse of a Function

Frances Van Dyke

February 1996

Edited by Timothy V. Craine, *Central Connecticut State University, New Britain, Connecticut;* Kim Girard, *Glasgow High School, Glasgow, Montana;* Guy R. Mauldin, *Science Hill High School, Johnson City, Tennessee;* Henri Picciotto, *The Urban School, San Francisco, California*

TEACHER'S GUIDE

Students have always found the topic of inverse functions difficult. When asked to find an inverse function algebraically, they often follow their first instinct, which is to take the reciprocal of whatever is given. This tendency is reinforced by the common use of the $f^{-1}(x)$ notation. If the function has taken an x to a y, we tell students that we want to take the y and find the process that will lead us back to the x. However, the idea of undoing what the function did gets lost in the mechanics of switching the x and y and then solving for the newly named y.

> **The idea of undoing can get lost in the process**

These activity sheets are designed to make the procedure for finding the inverse of a function easier and more meaningful. Sheet 1 gives students practice in thinking about undoing what was done in a variety of situations. It illustrates that undoing a sequence of steps requires proceeding in reverse order, first undoing what was last done. One-, two-, and three-step functions are considered. The sheet contains examples from everyday life as well as purely algebraic compositions. Arrows separate the steps; the nature of the step is denoted above the arrow.

In the words of the *Curriculum and Evaluation Standards for School Mathematics*, "[a] function can be described by a written statement, by an algebraic formula, as a table of input-output values, or by a graph" (NCTM 1989, 154). On sheets 2 through 4, the studen is given or asked to furnish each of these representations for a particular function. On the bottom half of each sheet, the student supplies the representations for the corresponding inverse function, again deriving the inverse by undoing the steps in the original function.

Graphing the two functions on the same set of axes should be avoided, since in most applications, different quantities are measured along the two axes and scales vary. For each example, the teacher can graph the function on a transparency and then flip it over the line that makes a 45-degree angle with the positive x- and y-axes. This procedure will correctly orient the axes to give a picture of the inverse function. The transparency can be passed around the room for the students to see. Alternatively, students can use heavy pens or pencils to make their own graphs, flip them over the 45-degree line, and hold them up to the light to see the image of the inverse function. See figure 1.

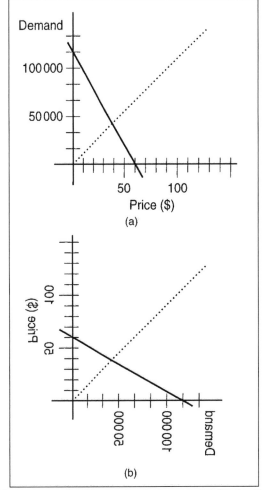

Fig. 1. the graph of the function is shown in (a); flipping the function, as shown in (b), pictures the inverse function.

Frances Van Dyke is an assistant professor of mathematical sciences at Central Connecticut State University, New Britain, Connecticut.

The sheets can be done by students working in cooperative groups right after the teacher has introduced the notion of inverse functions. Sheet 1 should be done first. In motivating students to complete this sheet, teachers may have them act out situations in which a multistep process can be reversed. For example, one student may act out this process: open the car door, get in the car, close the door, put key in ignition, and turn on ignition. A second student can then act out the reverse steps: turn off ignition, remove key from ignition, open the door, get out of the car, and close the car door.

Another physical approach to inverse operations involves a Rubik's cube (Garman 1985). Starting from the solved position, one student makes a move and a second student undoes it to return the cube to its original position. Students can then try examples involving two or more moves.

The teacher may also remind students that the process of undoing is used in solving equations. See Rubenstein et al. (1995, 105) for an example in which this process is illustrated with a flowchart.

Sheets 2 through 4 are independent of one another, although they increase in difficulty. Sheet 2 looks at the relationship between price and demand. The law of demand states that as the price decreases, the demand increases. Economists have traditionally graphed price as a function of demand, but to mathematicians, considering price as the independent variable seems more natural. Sheet 4 explores the relationship between exponential and logarithmic functions and should be used only if students have been introduced to e and natural logarithms.

> **Ask students to generate non-mathematical examples of inverse operations**

Each function on sheets 2 through 4 is presented in the context of an application in which variables other than x and y are used. If students are familiar with the $f(x)$ notation for functions, the inverse functional notation $f^{-1}(x)$ may be introduced as an extension. For example, on sheet 2, if the original function is written $d = f(p)$, then the inverse function is written $p = f^{-1}(d)$. This process leads to the generalization that if $y = f(x)$, then $x = f^{-1}(y)$. To retain y as the dependent variable, the latter is rewritten by switching x and y so that the inverse function is expressed as $y = f^{-1}(x)$.

After this activity has been completed, the traditional process of finding the inverse by switching the variables x and y and solving for y can be given as an alternative method. In each function for this activity, the independent variable appears only once in the functional expression. The method of switching variables is particularly helpful for such

functions as $f(x) = (2x + 3)/(x-4)$, in which the independent variable appears more than once. Finding the inverse of this function in the manner shown on the sheets requires dividing the denominator into the numerator and rewriting the function as $f(x) = 2 + 11/(x-4)$. Since this procedure is cumbersome, it is a good idea for students to learn the traditional method as well as the method presented herein.

Each function presented in this activity is a one-to-one function. Teachers should point out that only one-to-one functions have inverse. More advanced classes may discuss restricting the domain of a function such as $f(x) = x^2$ so that an inverse can be found.

Grade levels: 9–12

Materials: Activity sheets and transparencies for the graphs of each of the three functions given on sheets 2–4

Objectives: Students will understand the inverse of a function and how to find the inverse through the process of undoing.

Prerequisites: Students should be familiar with tabular, graphical, and algebraic representations of functions and have experience modeling real-world phenomena that are functions. The idea of an inverse function and the notion of a one-to-one function should have been discussed.

Directions: The sheets should be done in order. They are designed to increase in difficulty.

Assessment: Ask students to explain what is meant by the *inverse of a function*. Ask them to generate additional nonmathematical examples of inverse operations, like those on sheet 1 (e.g., the inverse of putting on socks then shoes is taking off shoes then socks). Ask students for the inverses of various elementary functions. Have students make up problems by composing elementary functions to form a function for which the rest of the class will find an inverse. Ask students to think of applications for which it is useful to find the inverse, that is, an application in which it is equally advantageous to express the independent variable in terms of the dependent variable.

Extensions: Have students think about processes that are their own inverses, such as finding the reciprocal of a nonzero number, reflecting in a mirror or over a line, rotating 180 degrees, or using a toggle command on a computer. Have students think about processes that do not have inverses because they cannot be undone, such as spilling a bottle of liquid on a sandy surface, creating nonrecyclable trash, or letting the water in a pan evaporate. A connection to the Second Law of Thermodynamics may be made.

BIBLIOGRAPHY

Garman, Brian. "Inverse Functions, Rubik's Cube, and Algebra." *Mathematics Teacher* 78 (January 1985): 33–34, 68.

National Council of Teachers of Mathematics. *Curriculum and Evaluation Standards for School Mathematics*. Reston, Va.: The Council, 1989.

Rubenstein, Rheta, et al. *Integrated Mathematics Book 1*. Boston: Houghton Mifflin Co., 1995.

Snapper, Ernst. "Inverse Functions and Their Derivatives." *American Mathematical Monthly* 97 (February 1990): 144–47.

Selected answers to sheets:

Sheet 1:

1. Cool down an object.

2. Subtract 10 from a number.

$$y \xrightarrow{\;-10\;} y - 10 = x.$$

3. Lose forty dollars.

4. Take the cube root. Using exponential notation will reinforce the idea that the inverse of taking a power is taking a root. On a calculator we can enter such examples as

$$(8\char`^3)\char`^(1/3)$$

as well as

$$8\char`^(1/3)\char`^3.$$

$$y \xrightarrow{\;(\;)^{1/3}\;} y^{1/3} = x.$$

5. Close a door.

6. Take the multiplicative inverse of a number. Here the function is its own inverse!

$$y \xrightarrow{\;(\;)^{-1}\;} y^{-1} = x.$$

7. Walk back from the end of the diving board and climb down the ladder.

8. Add 5 to the number and divide the result by 8.

$$y \xrightarrow{\;+5\;} y + 5 \xrightarrow{\;\div 8\;} \frac{y + 5}{8} = x.$$

9. Rewind the videotape and go outside.

10. Subtract 6 from a number and take the fifth root of the result.

$$y \xrightarrow{\;-6\;} y - 6 \xrightarrow{\;(\;)^{1/5}\;} \sqrt[5]{y - 6} = x$$

11. Take cargo off a boat, drive from the dock to the warehouse, and unload the truck.

12. Add 2000, divide the result by 62, and take the cube root.

$$x \xrightarrow{\;(\;)^{3}\;} x^3 \xrightarrow{\;*62\;} 62x^3 \xrightarrow{\;-2\,000\;} 62x\_2\,000 = y$$

$$y \xrightarrow{\;+2\,000\;} y + 2\,000 \xrightarrow{\;\div 62\;} \frac{y + 2\,000}{62} \xrightarrow{\;(\;)^{1/3}\;} \sqrt[3]{\frac{y + 2\,000}{62}} = x.$$

Sheet 2:

1.

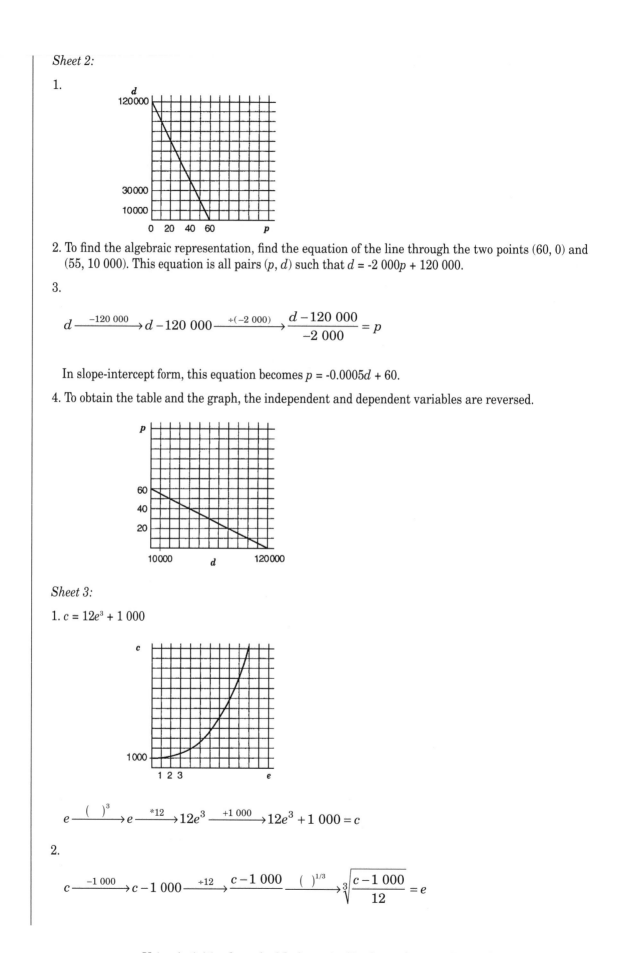

2. To find the algebraic representation, find the equation of the line through the two points (60, 0) and (55, 10 000). This equation is all pairs (p, d) such that $d = -2\,000p + 120\,000$.

3.

$$d \xrightarrow{\ -120\,000\ } d - 120\,000 \xrightarrow{\ \div(-2\,000)\ } \frac{d - 120\,000}{-2\,000} = p$$

In slope-intercept form, this equation becomes $p = -0.0005d + 60$.

4. To obtain the table and the graph, the independent and dependent variables are reversed.

Sheet 3:

1. $c = 12e^3 + 1\,000$

$$e \xrightarrow{\ (\ )^3\ } e \xrightarrow{\ *12\ } 12e^3 \xrightarrow{\ +1\,000\ } 12e^3 + 1\,000 = c$$

2.

$$c \xrightarrow{\ -1\,000\ } c - 1\,000 \xrightarrow{\ \div 12\ } \frac{c - 1\,000}{} \xrightarrow{\ (\ )^{1/3}\ } \sqrt[3]{\frac{c - 1\,000}{12}} = e$$

Using Activities from the *Mathematics Teacher* to Support *Principles and Standards*

3.

$$e = \sqrt[3]{\dfrac{c - 1\,000}{12}}$$

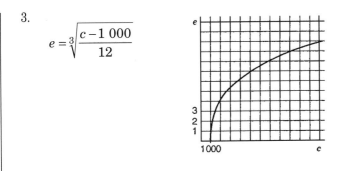

Sheet 4:

1. Answers will vary.

2. When $t = 1\,620$ is substituted into $A = 60e^{-0.0004279t}$, the result is $A \approx 30$ grams, or about half the original amount. Thus t is the half-life of the substance. Students may encounter the equivalent formula $A = 60(0.5)^{t/1\,620}$ when studying about half-life in a science class.

3.

$$t \xrightarrow{\ *0.0004279\ } 0.0004279t \xrightarrow{\ \ln\ } e^{-0.0004279t} \xrightarrow{\ *60\ } 60e^{-0.0004279t} = A$$

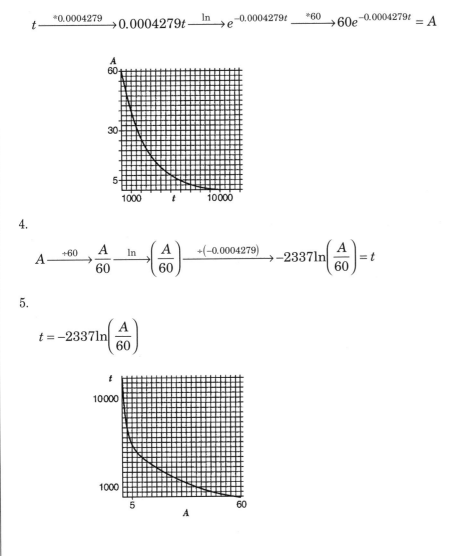

4.

$$A \xrightarrow{\ \div 60\ } \dfrac{A}{60} \xrightarrow{\ \ln\ } \left(\dfrac{A}{60}\right) \xrightarrow{\ \div(-0.0004279)\ } -2337\ln\left(\dfrac{A}{60}\right) = t$$

5.

$$t = -2337\ln\left(\dfrac{A}{60}\right)$$

Write a phrase or sentence that undoes the action given and brings it back to the original state. Fill in the arrow sequences where indicated.

Action Taken	Inverse Operation

1. Heat up an object.

2. Add 10 to a number.

$$x \xrightarrow{+10} x + 10 = y \qquad\qquad y \xrightarrow{} = x$$

3. Win forty dollars.

4. Cube a number.

$$x \xrightarrow{(\ )^3} x^3 = y \qquad\qquad y \xrightarrow{} = x$$

5. Open a door.

6. Take the multiplicative inverse of a number.

$$x \xrightarrow{(\ )^1} x^{-1} = y \qquad\qquad y \xrightarrow{} = x$$

7. Climb up the ladder and walk out to the end of the diving board.

8. Multiply a number by 8 and subtract 5 from the result.

$$x \xrightarrow{\bullet 8} 8x \xrightarrow{-5} 8x - 5 = y \qquad y \xrightarrow{} \xrightarrow{} = x$$

9. Go inside and fast-forward a videotape.

10. Take a number to the fifth power and add 6 to the result.

$$x \xrightarrow{(\ )^5} x^5 \xrightarrow{+6} x^5 + 6 = y \qquad y \xrightarrow{} \xrightarrow{} = x$$

11. Load a truck at a warehouse, drive to a dock, and put cargo on a boat.

12. Cube a number, multiply the result by 62, and subtract 2000.

$$x \xrightarrow{} \xrightarrow{} \xrightarrow{} = y$$

$$y \xrightarrow{} \xrightarrow{} \xrightarrow{} = x$$

The Best Value hardware-store chain has produced a new wonder shovel. Market research shows that if the price is $60, no one will buy it. For each $5 drop in price, however, an additional 10000 customers will buy the shovel. Thus demand is a linear function of price. Three representations for this function follow.

1. Add three more pairs to the table of values. Graph the function.

2. Explain how the algebraic expression for the function was obtained.

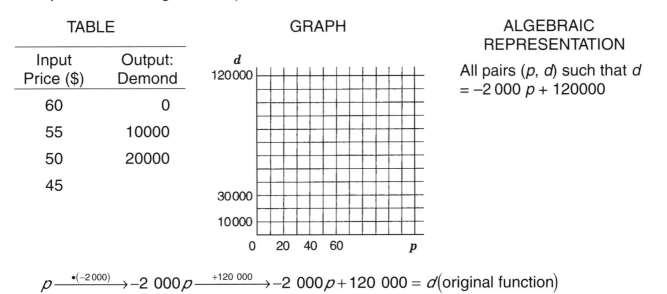

TABLE		GRAPH	ALGEBRAIC REPRESENTATION

TABLE

Input Price ($)	Output: Demond
60	0
55	10000
50	20000
45	

ALGEBRAIC REPRESENTATION

All pairs (p, d) such that $d = -2\,000\,p + 120000$

$$p \xrightarrow{\bullet(-2\,000)} -2\,000p \xrightarrow{+120\,000} -2\,000p + 120\,000 = d \text{(original function)}$$

The marketing department at Best Value wants to know how to find the price that it should charge to sell all the shovels it has in inventory (the demand).

3. Complete the sequence of arrows to find the inverse function.

$$d \longrightarrow \qquad \longrightarrow \qquad = p \text{(inverse function)}$$

4. Give the inverse function's three representations.

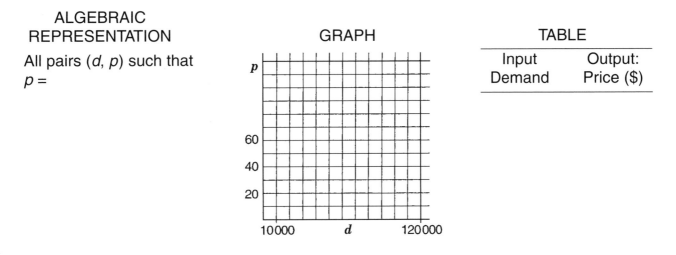

ALGEBRAIC REPRESENTATION

All pairs (d, p) such that $p =$

TABLE

Input Demand	Output: Price ($)

From the *Mathematics Teacher*, February 1996

A sculptor wants to build a magnificent cube for an exhibition. His fixed costs are $1 000, and he calculates that he will need to spend $12 per cubic meter for the materials. The cost in dollars is a function of the length in meters of the edge of the cube.

1. Give three representations for the function.

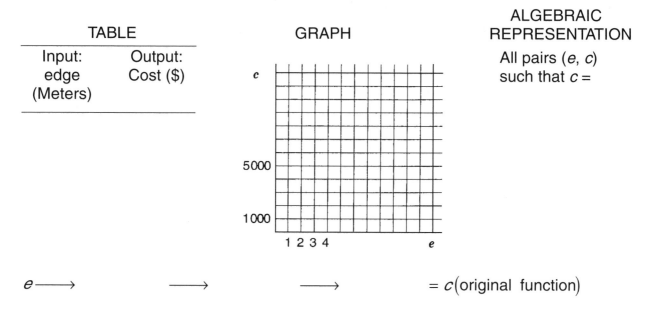

TABLE

Input: edge (Meters)	Output: Cost ($)

GRAPH

ALGEBRAIC REPRESENTATION

All pairs (*e*, *c*) such that *c* =

$e \longrightarrow \qquad \longrightarrow \qquad \longrightarrow \qquad = c$(original function)

The sculptor may need to use the inverse function to let him know how big a cube he can build for a certain amount of money.

2. Complete the sequence of arrows to find the inverse function.

$c \longrightarrow \qquad \longrightarrow \qquad \longrightarrow \qquad = e$(inverse function)

3. Give three representations for the functions.

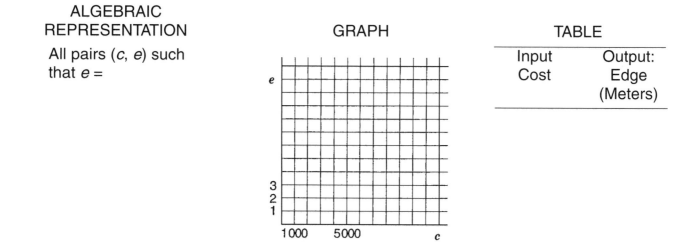

ALGEBRAIC REPRESENTATION

All pairs (*c*, *e*) such that *e* =

GRAPH

TABLE

Input Cost	Output: Edge (Meters)

From the *Mathematics Teacher*, February 1996

Radium, Ra^{226}, is a radioactive substance that decays. If 60 grams are stored in a container, the amount A left after t years is approximately $A = 60e^{-0.0004279t}$.

1. Continue the table with at least three more pairs of values.

2. Explain why the half life of this substance is 1 620 years.

3. Show the graphical representation.

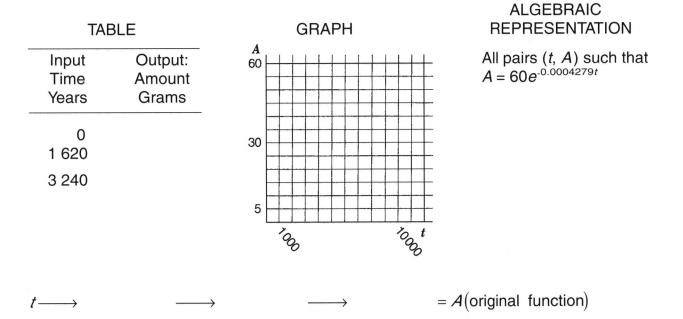

| TABLE | | GRAPH | ALGEBRAIC REPRESENTATION |

TABLE

Input Time Years	Output: Amount Grams
0	
1 620	
3 240	

GRAPH

ALGEBRAIC REPRESENTATION

All pairs (t, A) such that $A = 60e^{-0.0004279t}$

$t \longrightarrow \qquad \longrightarrow \qquad \longrightarrow \qquad = A(\text{original function})$

Scientists need to know how much time will pass until only such a small amount of the Ra^{226} is left that it poses no safety hazard.

4. Complete the sequence of arrows to find the inverse function.

$A \longrightarrow \qquad \longrightarrow \qquad \longrightarrow \qquad = t(\text{inverse function})$

5. Give three representations for the fuonction.

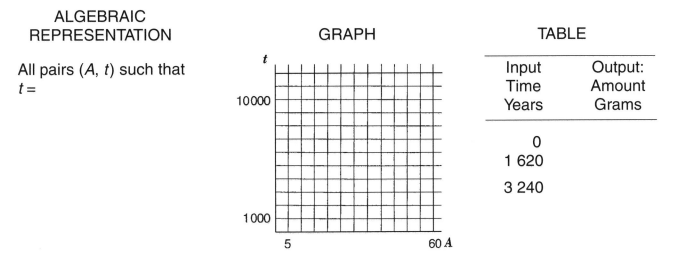

ALGEBRAIC REPRESENTATION

All pairs (A, t) such that $t =$

GRAPH

TABLE

Input Time Years	Output: Amount Grams
0	
1 620	
3 240	

Experiments from Psychology and Neurology

William S. Hadley

October 1996

Edited by Henri Picciotto, *The Urban School, San Francisco, California.*

TEACHER'S GUIDE

One topic advocated by the new curriculum-reform movement, but often missing from textbooks, is data collection and analysis. As a classroom teacher, I have tried many different experiments in an attempt to produce good linear-data sets: measuring different body parts, burning candles, and so on. Unfortunately, students were often unable to make accurate enough measurements to obtain usable data, and many experiments required students to collect data over a long time or outside of class. As a result, I have often been disappointed.

In contrast, the data for each of the following experiments can be easily collected in one forty-minute class period, and the results are fairly easily interpreted by the students. These experiments have been successfully used by the author with adults as well as with mathematics classes ranging from general mathematics to calculus, in urban and suburban schools. Students have consistently found these experiments to be enjoyable and have always reacted enthusiastically. They are also currently part of the data-analysis unit in the PUMP (Pittsburgh Urban Mathematics Project) first-year-algebra course.

> *The data can be easily collected*

Objectives: The mathematical heart of the activity comprises—

- the interpretation of the linear equations obtained from the data;
- their use in making predictions, including interpolating and extrapolating; and
- their use in comparing related data sets.

The skills that are developed and reinforced are—

- finding a linear equation from the coordinates of points and
- manipulating linear equations.

Prerequisites and grade level: Students should have some basis understanding of graphing and of linear equations and their graphs. These activities have been used in general-mathematics, first- and second-year-algebra, precalculus, and calculus classes.

Materials: For both sets of experiments, stopwatches (coaches and science teachers are good sources), graph paper, and clear plastic rulers are needed. In addition, an overhead projector is required for the Stroop experiments.

Directions: The activity comprises two unrelated sets of experiments. Sheets 1 and 2, the Stroop experiments from cognitive psychology, can be completed independently of sheets 3 and 4, the human-chain experiments from neurology, or vice versa. Or all can be done. In each situation, students perform two similar experiments involving the timing of certain activities. Each experiment yields a set of data points that are well approximated by a linear function. Become familiar with the activity sheets before reading the subsequent comments.

Data collection

The Stroop test can be humorous and enjoyable for most subjects, but some students may be color-blind or too anxious to participate. Similarly, in the human-chain experiments, some students may have cultural problems touching other students or being touched. Those students can be assigned to be timers or data recorders. In this way, at least four or five students can still participate in the data

---

Editorial Comment: For this publication the matching and nonmatching lists could not be printed in color. To create your nonmatching list, the same ink color is not always used for each nonmatching word. For example, GREEN may be written in red ink at one place in the list and in black ink at another. BLACK and RED may appear consecutively but both be written in blue ink.

Bill Hadley has been a teacher for the past twenty-five years in the Pittsburgh public school system and is currently working with Carnegie Mellon University to create the PUMP algebra curriculum and intelligent computer tutors.

63

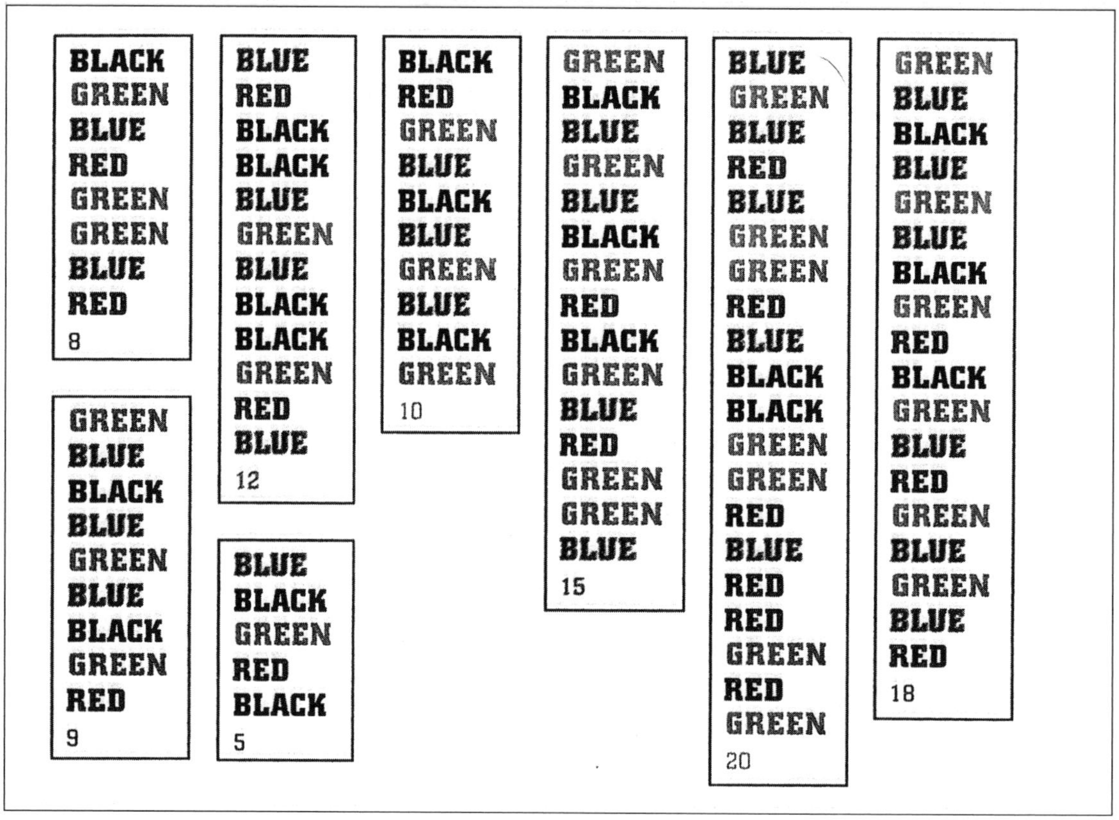

Fig. 1. Matching List

collection without being involved in the part of it that might cause them discomfort.

When students have found the average of the three data values for each experiment, teachers may want to use the median rather than the mean. In fact, this choice coulde discussed with the class.

Data analysis

Since the experiments yield data that are very close to linear, it is not difficult to find a reasonably good line to fit the data by approximating it with a ruler or a piece of uncooked spaghetti. A statistician's techniques for fitting a line, such as the median-median line or linear regression, are not necessary at this level. Instead, students can eyeball the line and find its equation by using algebraic manipulation or by observing the graph. It is preferable to use the slope-intercept form because those parameters will be discussed in the questions that follow.

In some instances, graphing-calculator technology can be used to fit the line. Such calculators as the TI-82, which permit multiple plots, are especially useful because they allow the simultaneous display of both data sets and allow fitting the line by regression or by the median-median approach.

Moreover, such calculators allow interpolation and extrapolation in several different ways: tracing a graph, inspecting a table of values, or directly evaluating the function.

Question 13 on sheets 2 and 4 asks the students to extrapolate from the data obtained. The purpose is to spark a discussion of what sort of extrapolation is legitimate, how much it can be trusted, and at what point it becomes absurd.

Sheets 1 and 2: The Stroop experiment: The experiment is described on sheet 1. We usually run this experiment by making the matching and non-matching lists on an overhead transparency. See figures 1 and 2 for sample lists. Our original versions were made with an ink-jet printer, but a copy store can make colored transparencies from the masters for a couple of dollars. They can also be made with blank transparencies and overhead-projector pens.

We cut each transparency into individual lists. The length of each list is best kept from five through twenty words. Making the lists for both experiments the same length is good practice. Using a couple of the lists more than once is also appropriate so as to have multiple points for the

Using Activities from the *Mathematics Teacher* to Support *Principles and Standards*

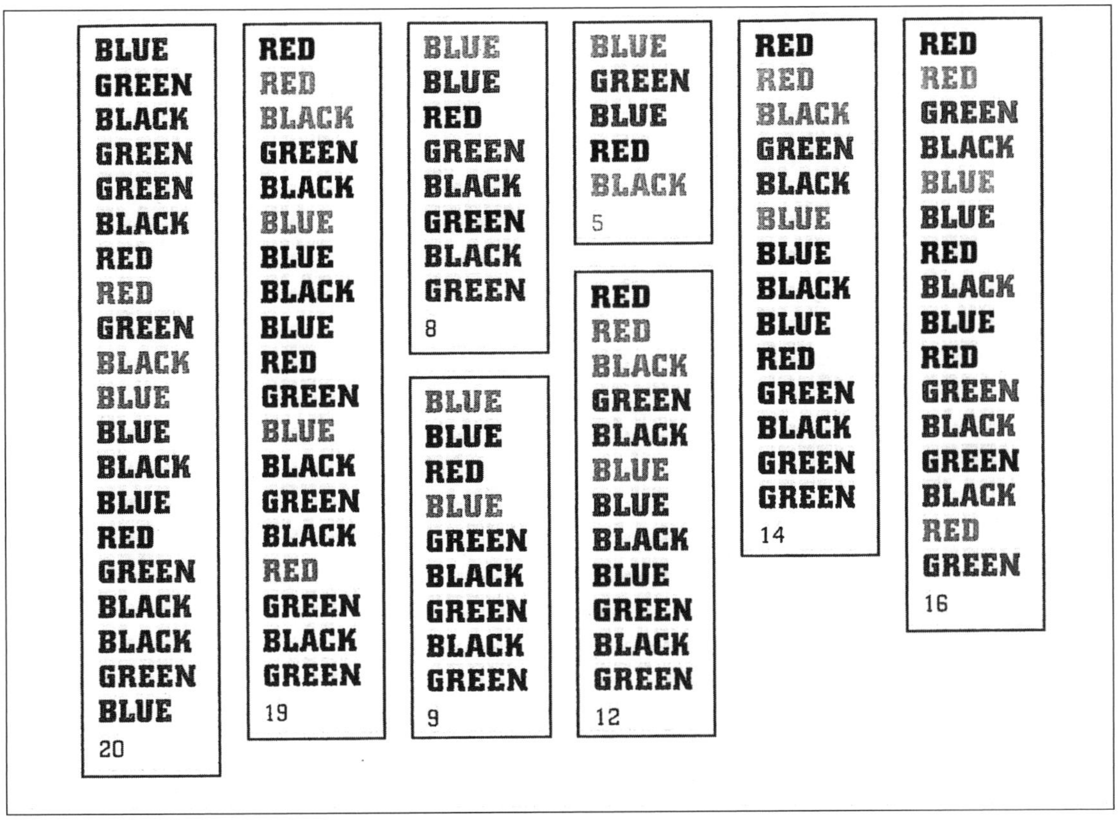

BLUE
GREEN
BLACK
GREEN
GREEN
BLACK
RED
RED
GREEN
BLACK
BLUE
BLUE
BLACK
BLUE
RED
GREEN
BLACK
BLACK
GREEN
BLUE
20

RED
RED
BLACK
GREEN
BLACK
BLUE
BLUE
BLACK
BLUE
RED
GREEN
BLUE
BLACK
GREEN
BLACK
RED
GREEN
BLACK
GREEN
19

BLUE
BLUE
RED
GREEN
BLACK
GREEN
GREEN
8

BLUE
BLUE
RED
BLUE
GREEN
BLACK
GREEN
BLACK
GREEN
9

BLUE
GREEN
BLUE
RED
BLACK
5

RED
RED
BLACK
GREEN
BLACK
BLUE
BLUE
BLACK
BLUE
GREEN
BLACK
GREEN
12

RED
RED
BLACK
GREEN
BLACK
BLUE
BLUE
BLACK
BLUE
RED
GREEN
BLACK
GREEN
GREEN
14

RED
RED
GREEN
BLACK
BLUE
BLUE
RED
BLACK
BLUE
RED
GREEN
BLACK
GREEN
BLACK
RED
GREEN
16

Fig. 2. Nonmatching List

same list. Because results are fairly consistent, the same student need not tackle matching and non–matching lists.

A helpful approach is to have the class discuss this experiment, cognitive psychology, and experimental procedure before actually collecting the data. Start with the questions on sheet 1, and add the following:

- What differences do we expect for the two types of lists? Why?

- Why would a cognitive psychologist be interested in these experiments?

After collecting and analyzing the data, another discussion should ensue about questions 14 and 15. See the subsequent suggestions for extensions. Depending on the level of the class, the students could be given these questions to answer first on their own or in groups; alternatively, the questions could be addressed in a whole-group discussion.

To run the experiment, the teacher places one list on the turned-off overhead projector without letting the subject know whether the list is matching or nonmatching. The teacher then makes sure that the timers and the individual who will name the colors are ready and turns on the projector. The student will immediately state the color of the ink for each word as quickly as possible. A mistake should be corrected before the student moves on. The timers keep time from when the overhead projector is turned on until the student has successfully completed the list. At least one dry run is helpful to make sure that all the students understand what is expected.

We normally perform the experiment ten times with the matching and ten times with the nonmatching lists, trying to order them randomly so not to give a student the advantage of knowing which type of list she or he will be reading. It it also a good idea to have at least one list of each type be read by two or more students to show variation. This process usually gives enough data to do the analysis, but performing it more times so that every student gets to do it once ensures that no one is left out. With three timers; three recorders, one for the overhead transparency or chalkboard for each list and another to calculate the average time; and twenty lists to be read by different students, this experiment almost ensures total participation.

Sometimes students become so flustered that they take an inordinate amount of time to complete their list. This situation gives the teacher an excellent opportunity to discuss the concept of an outlier and to have the students during their analysis discuss the effect of such a data point on the analysis.

Sheets 3 and 4: Human-chain experiments: This experiment is straightforward. Choose the lengths of the chains on the basis of the number of students in the class. However, larger numbers of "links" can be obtained by making a circle and passing the pulse through more than one cycle. Having students close their eyes improves the accuracy of the results; otherwise, the students can actually "see" the pulse coming.

The main thing that makes the chain experiments so interesting is the difference in the slopes of the two equations. The y-intercepts should be and usually are virtually identical, but differences in the slopes are between 0.08 and 0.15 seconds per person. This difference is the average time the nerve impulse takes to travel up one's arm. In a high school classroom, the distance from a student's wrist to the shoulder is approximately two feet. Students can calculate this distance more accurately by measuring it for each student and finding the average. Using this difference, students can estimate the actual nerve speed to be approximately 0.05 to 0.12 seconds per foot, which is about 20 feet per second, 1200 feet per minute, or 15 miles per hour.

Assessment: The "conclusions" questions on the analysis sheets afford an opportunity for assessing students' understanding.

Extensions: Students can create their own experiments involving the timing of an activity. One that comes to mind is exploring whether one gets better at the Stroop test through practice. A similar experiment to the human chain is doing "the wave." Students will have their own ideas.

Answers: Sheet 1: See Teacher's Guide.

Sheet 2: Most answers depend on the data collected.
1–3. Answers will vary.
 4. Seconds
 5. Start-up time, reaction time of timers. Sometimes the y-intercept will be negative because of an outlier, which affords a good opportunity to discuss whether to include this data point.
 6. Seconds per color named
 7. Time per word, or increase in time for each word added to list
8–10. Answers will vary.

11. This form permits easier predictions about the number of colors identified in a given time.
12. Words per second
13. Answers will vary. The numbers obtained by these types of calculations cannot be really trusted, since even after an hour, the subject may slow down because of fatigue or speed up because of the additional practice. It is impossible even to conceive of someone's doing this activity for a day or a year.
14. Answers will vary. One student has answered, "Nonmatching takes more time because you have to make your brain read the color of the ink, not the word."
15. Answers will vary.

Sheet 3: See Teacher's Guide.

Sheet 4: Most answers depend on the data collected.
1–3. Answers will vary.
 4. Seconds
 5. Start-up time, reaction time of timers
 6. Seconds per person
 7. Time added to the experiment for each additional person in the chain
8–10. Answers will vary.
11. It is easier to make predictions about the number of people through whom the impulse can go in a given time.
12. People per second
13. Answers will vary. The logistics of doing the experiment on this scale seem unsurmountable.
14. One could make longer chains by going around more than once, but people's ability to concentrate is limited. After an hour, it is unlikely that the pulse would still be going, that no additional pulse would have started, and so on.
15. Answers will vary.
16–17. See Teacher's Guide.

BIBLIOGRAPHY

Stroop, J. R. "Studies in Interference in Serial Verbal Relations." *Journal of Experimental Psychology* 18 (1935): 643–62.

Winter, Mary Jean, and Ron Carlson. *Algebra Experiments I.* Menlo Park, Calif.: Addison-Wesley Publishing Co., 1993.

Using Activities from the *Mathematics Teacher* to Support *Principles and Standards*

One of the main uses of data analysis is to make predictions about real-world situations. The first step is to collect the data. We are going to perform an experiment from cognitive psychology, which is the branch of psychology that tries to understand and explain how the human brain works. The experiment is named after the man who first performed it, J. R. Stroop.

- Each student will look at a list of color words—red, green, black, or blue. Each list is a different length. Each word will be written in red, green, black, or blue ink.
- The student will be asked to say the color of the ink for each word. The time needed to complete each list will be recorded.
- Two different lists will be used: one on which the color of the ink matches the color word, for example, red written in red ink, and a second on which the color of the ink does not match the color word, for example, red written in blue ink.
- The first type of list is called matching, and the second is called nomatching. The length of the list will be recorded, and three different students will record the time a fourth student takes to name the color of the ink of each word on the list. The student's average time will be calculated.

1. What do you think that we will find when we perform these experiments?
2. What questions should we be trying to answer?
3. Why do we use three timers? How can we combine the three measurements?

Matching Experiment

List Length	Timer 1	Timer 2	Timer 3	Average Time

Nonmatching Experiment

List Length	Timer 1	Timer 2	Timer 3	Average Time

1. Graphing: Construct a graph that displays the points for the matching experiment using dots for your points. Graph the length of the list on the horizontal axis and the time on the vertial axis. On the same grid, plot the points for the nonmatching experiment using small squares.

2. Draw a line for the matching experiment by moving a clear plastic straightedge until you believe that you have the line that best fits these points. It should pass as close to the data points as possible, with some points above and others below. Repeat for the non-matching experiment.

Answer questions 3–13 for both the matching experiment and the nonmatching experiment.

3. Finding equations: For your lines, find equations, in slope–intercept form.

   $Y =$ _____ $X +$ _____   (matching)

   $Y =$ _____ $X +$ _____   (nonmatching)

4. What are the Y–intercepts of your lines? In what unit are they measured?

5. What do the Y–intercepts mean in these experiments? Explain.

6. What are the slopes of your lines? In what unit are they measured?

7. What do the slopes mean in these experiments? Explain.

8. Making predictions: Using your graphs or your equations, estimate how long it would take to name the colors in a list of twenty-five words. How long would it take for ten words? Explain

9. Using your equations, estimate for how many words you could name the color in two minutes. Explain.

10. Solve your equations for X:

   $X =$ _____ $Y +$ _____   (matching)

   $X =$ _____ $Y +$ _____   (nonmatching)

11. What is the purpose of having the equations in this form?

12. What are the unit and the meaning of the slope in these equations?

13. Using your equations, estimate for how many words you could name the color in one hour, in one day, and in one year. Explain how you found these answers and what your assumptions were. Do you trust the answers you obtained? Why or why not?

14. Conclusions: Are the results of the matching and nonmatching experiments what you had expected? Comment on the differences and similarities between the two experiments.

15. Suppose that you were the cognitive psychologist who designed this experiment. Write up your conclusions.

The human-chain wrist experiment is performed by making a human chain. This chain is formed by holding the wrist of the person to your right. Your teacher will pick a student to begin the chain and another to end it. The members of the chain must keep their eyes closed. When the teacher says "Go!" the first student in the chain should gently squeeze the wrist of the student to the right, who then squeezes the next person's wrist, and so on, until the last person in the chain feels the squeeze. The last student should then say "Stop!" Three timers will keep the time from when the teacher says go until the last student says stop.

The experiment will be repeated ten times, with longer and longer chains of students.

Human–Chain Wrist Experiment

Number of Students	Timer 1	Timer 2	Timer 3	Average Time

The human–chain shoulder experiment is performed in the same way, except that each person holds the shoulder of the person to the right.

Human–Chain Shoulder Experiment

Number of Students	Timer 1	Timer 2	Timer 3	Average Time

1. What do you think that we will find when we perform these experiments?

2. What questions should we be trying to answer?

3. Why do we use three timers? How can we combine the three measurements?

1. Graphing: Construct a graph of your data with the number of students on the horizontal axis and the time on the vertical axis. Use dots for the wrist experiment and small squares for the shoulder experiment.

2. Draw a line for the wrist experiment by moving a clear plastic straightedge until you believe that you have the line that best fits these points. It should pass as close to the data points as possible, with some points above and others below. Repeat for the shoulder experiment.

Answer questions 3–10 for both the wrist experiment and the shoulder experiment.

3. Finding equations: For your lines, find equations in slope–intercept form.

$Y =$ _____ $X +$ _____ (wrist)

$Y =$ _____ $X +$ _____ (shoulder)

4. What are the Y-intercepts of your lines? In what unit are they measured?

5. What do the Y-intercepts mean in these experiments? Explain.

6. What are the slopes of your lines? In what unit are they measured?

7. What do the slopes mean in these experiments? Explain.

8. Making predictions: Using your graphs or your equations, estimate how long it would take to pass this squeeze through a chain of 50 people. How long would it take for 100 people? For 10 000 people? For the entire world's population of 5 billion people?

9. Using your equations, estimate through how many people the squeeze could pass in an hour.

10. Solve your equations for X:

$X =$ _____ $Y +$ _____ (wrist)

$X=$ _____ $Y +$ _____ (shoulder)

11. What is the purpose of having the equations in this form?

12. What are the unit and the meaning of the slope in these equations?

13. Using your equations, estimate through how many people the squeeze could pass in one day and in one year. Explain how you found these answers and what your assumptions were. Do you trust the answers you obtained? Why or why not?

14. Could we test any of these predictions in the classroom? If yes, how? If no, why not?

15. Conclusions: Are the results of the wrist and shoulder experiments what you had expected? Comment on the differences and similarities between the two experiments.

16. How do you interpret the difference between the slopes?

17. How fast do nerve impulses travel through one person's body?

Packing Them In

Claudia R. Carter   March 1997

Edited by Henri Picciotto, *The Urban School, San Francisco, California*

TEACHER'S GUIDE

Spheres and cylinders are usually packaged in rectangular boxes. Is this packaging the best choice? In this activity, students investigate some questions in the mathematics of packaging, and in the process they explore and use various concepts in geometry, including area and the Pythagorean theorem. The mathematics comes out of the discussion of the packaging of soft-drink cans into six-packs and focuses on cost-effectiveness in terms of the horizontal area used and on the need for the packages to be easy to juxtapose on a shelf. A side benefit of the activity is to help students realize how the knowledge of mathematics allows exact answers that lead to generalizations and formulas.

Grade levels: 8–12

Materials: Twelve soft-drink cans for each group; calculators; a compass, a protractor, a ruler; geometric templates—a plastic sheet with different geometric shapes cut out—or some circular object that can be traced onto graph paper; dynamic-geometry software if students have access to computers. For sheet 3, students need chart or poster paper.

Prerequisites: Students should know what area is and how to calculate the areas of circles, rectangles, and triangles. Students should have had some experience with equilateral triangles and 30-60-90 triangles, or at least with the Pythagorean theorem.

Preliminary discussion: Prior to handing out the first activity sheet, you may hold a brief class discussion of the main question that students will be investigating: Is a rectangular six-pack the best configuration for soft-drink cans? It is necessary to decide on a definition of *best*, since the "best" package can be interpreted in several ways. Although other issues may come up and students should be encouraged to explore them, the definition of best used in sheets 1–3 concerns how efficiently the horizontal space is used for the cans in the package. For more on this question, see unit 12 of *The Right Stuff* (COMAP forthcoming), which inspired this activity. In particular, that unit addresses the question of why we use ratios rather than differences to discuss how effectively the space is used: ratios can be compared independently of the measure of the radius of the circle or the number of cans in the package. Differences, however, are dependent on the sizes and numbers of the objects and hence cannot easily be used for comparison.

Sheet 1: The purpose of this sheet is to define efficiency and to calculate the efficiency of a standard, rectangular six-pack. The students should be able to answer the questions by reasoning rather than measuring. In other words, they should see six radii across the can, so the total length of the box would be 18 cm when the radius of the cans is 3 cm. You might want to use this opportunity for a discussion about points of tangency and other geometric properties that are pertinent to the diagrams involved. Note that the actual radius of a standard soft-drink can is slightly greater than 3 cm.

Students might conjecture that the efficiency of the container should increase with the number of circles or with the radius of the circle. In the exercises, students discover that the container grows in proportion to the number of circles or to the measure of their radii. Since we defined efficiency as a ratio, it is unaffected by the number or size of the circles. This surprising result is the object of problems 5–8.

Sheet 2: This sheet looks at a specific arrangement of cans to introduce ideas for use in sheet 3.

The triangle is equilateral, since its sides are all $2r$. Since the sides of the parallelogram are parallel to the sides of the triangle, it follows that the angles of the parallelogram have measures of 60 degrees and 120 degrees. The height of the equilateral triangle is $r\sqrt{3}$, so the height of the parallelogram is equal to $2r + r\sqrt{3}$. Similarly, the base is the sum

$$\frac{r}{\sqrt{3}} + 4r + r\sqrt{3} \ .$$

Claudia Carter teaches at the Mississippi School, Columbus, Mississippi. She is interested in improving the teaching of mathematics through real-world applications, the integration of topics and disciplines, and technology.

Make sure that students understand that measuring the height and base will not give them as precise an answer as does the foregoing method, which is absolutely accurate for any value of *r*. Students who choose to trace the circles and then measure the resulting parallelogram need to realize that the method is subject to human error and that they are approximating the answers. Moreover, students have a tendency to think that their figures are more accurate when they use the computer. This situation is true only if their constructions are obtained by a correct use of points of tangency, parallel and perpendicular lines, and so on. However, if their figures are drawings rather than correct constructions, then their data are estimations. In dynamic-geometry software, such as Cabri Geometry II (Bellemain and Laborde 1995) or The Geometer's Sketchpad 3 (Jackiw 1995), a correct construction will retain all its geometric properties if any of its points are dragged.

For further insights into the comparison of staggering versus stacking and just how much efficiency one can expect, see *The Right Stuff* (COMAP forthcoming).

Sheet 3: This activity is more open ended, so it is difficult to predict exactly what students will do. In my experience, students have worked with shapes ranging from parallelograms to hexagons, most of which have been regular. In addition, some students have taken all three dimensions into account in their investigations. For example, one group proposed to stack three cans on the bottom and three on the top using an equilateral triangle for a base. Another group created a cylinder to hold the cylindrical cans. Probably the most common shape that students use is the hexa-pack containing seven cans.

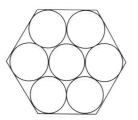

Students should work in teams of four. You may want to allow two class periods for the exploration, especially if students can go to a computer laboratory, because dynamic-geometry software may support more avenues of exploration. In either instance, students can do oral presentations on the next day, or if you are pressed for time, you can ask for the presentations to be in the form of posters.

Be prepared to help teams that want to calculate the area of an unfamiliar shape, perhaps by making books or formulas available or by suggesting techniques that break up the figure into familiar shapes.

One unanticipated outcome of this activity was that some of the groups that chose containers of the same shape did not obtain the same efficiency ratings. This discovery reinforced the need to use mathematics to obtain exact answers rather than to use measurement for approximate answers.

Assessment of sheet 3: Using a rubric for grading the presentations or posters is helpful to both the teacher and the teams. You should make teams aware of the rubric. You may choose to have students participate in the grading process by self-evaluations or by having a member from each group evaluate the other teams. A suggested rubric follows: Using a scale from 1 through 5, where 5 is best, grade each category.

1. Team effort

   • Did all the team members participate?

   • Did the members work well together?

2. Knowledge and understanding of mathematics

   • Were appropriate mathematics formulas used?

   • Were the answers reasonable?

   • Did the team members use appropriate mathematics terminology?

3. Use of visual aids

   • Was the choice of visual aids appropriate?

   • Were the visual aids created accurately?

   • Did the visual aids clarify and promote the audience's understanding?

4. How well the team addressed "best" argument

   • Did the team clearly state its choice and the reasons for the decision?

   • Did the team use efficiency as the defined ratio?

   • If the team also used other criteria, did it explain them clearly?

   • Did the team explain the ideas and calculations that supported its argument?

Sheet 4: Sheet 4 provides the opportunity for students to analyze some of the shapes that their team may not have explored on sheet 3. In addition, it explores the question of whether the packages tessellate the plane, and therefore the grocery shelf or the floor of a truck.

Extensions: You may want to keep the students' designs for future reference. They make great display material for the classroom. The ideas could be expanded into projects that look more closely at particular geometric shapes or that investigate other aspects of packaging, such as the cost of materials, surface areas, limitations on the actual weight or volume, how efficiency could be defined if the package were not a prism, and so on. Another question of efficiency and tiling that students can explore is how much cardboard is wasted when the pattern for the box is cut out of larger sheets of cardboard.

Solutions:
Sheet 1:
1. 18 cm by 12 cm
2. With the ratio, we represent the part (the area of the bottoms of the cans) to the whole (the area of the container). The greater the ratio, the less space that is wasted. Ratios allow us to convert easily to percents, which give us a convenient way to discuss efficiency.
3. Exact efficiency = $\pi/4$; approximate efficiency = 0.785.
4. 78.5 percent
5. $\pi/4$. If we view the six-pack as made up of six one-packs, we see that the area of the box is multiplied by 6, as is the area of the can. Since we multiplied the numerator and the denominator by the same number, we get the same ratio.
6. We would expect the efficiency to be the same. See the answer to question 5.
7. The efficiency is the same for cans of any radius.
8. The radius squared is a common factor of the numerator and denominator, so the fraction can be simplified:

$$\frac{6\pi r^2}{6r \bullet 4r} = \frac{\pi}{4}$$

The efficiency would be 78.5 percent for cans of any radius.

Sheet 2:
1.

$$\left(\frac{12+20}{\sqrt{3}}\right)r^2 = 23.55r^2$$

2.

$$\left(\frac{27-15\sqrt{3}\pi}{4}\right) \approx 0.8005 = 80.05\%$$

3. The efficiency of the parallelo-pack is a little better than that of the standard six-pack.
4. It appears that staggered positioning may allow greater efficiency. If you allowed only stacking for the parallelogram, the efficiency would be poor, since a large area would not be covered. For the same reason, staggering would not be the most efficient method for the rectangular six-pack.
5. Answers will vary. Some possibilities: How well would it fit on a shelf or pack in a crate? On a shelf, would it be visible to the consumer? Would the labels be legible and appealing to the consumer? Would it be easy to carry and not have sharp edges?

Sheet 3: Answers will vary.

Sheet 4:
1. (a) 64.3 percent; (b) 72.9 percent; (c) 78.5 percent; (d) 79.3 percent; (e) 85.1 percent; (f) 77.8 percent
2. All the figures except the circle can tessellate the plane.
3. Although other shapes do tessellate, the rectangle is the easiest to work with in this regard. It allows for visibility and easy access. It is easy to carry and does not have sharp edges. It slides off the shelf without disturbing other packages. It may be easier to manufacture. You or your students may have other ideas.

REFERENCES

Bellemain, Franck, and Jean-Marie Laborde. Cabri Geometry II. Dallas, Tex.: Texas Instruments, 1995. Software.
Consortium for Mathematics and Its Applications (COMAP). *The Right Stuff.* ARISE project, unit 12. Lexington, Mass.: COMAP, forthcoming.
Jackiw, Nicholas. The Geometer's Sketchpad 3. Berkeley, Calif.: Key Curriculum

ARISE (Applications Reform in Secondary Education) is a comprehensive curriculum-development project funded by a National Science Foundation grant to the Consortium for Mathematics and Its Applications (COMAP). This project has developed a core curriculum for grades 9 through 11 in which the mathematics to be studied arises out of applications.

You are familiar with soft drinks packaged in sixes. How efficiently does this arrangement use shelf space? Do other arrangements fit more cans on a shelf? In this activity you will explore these ideas. For starters, consider six cans packed in rows in a rectangular package, the standard six-pack. The following figure shows a two-dimensional view of it, with each circle representing the bottom of a can. Use the following questions to help you determine how efficient the package is. Be sure to justify your answers and explain how you found them.

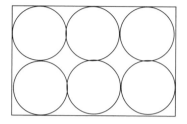

1. What are the dimensions of the bottom of the box if the radius is 3 cm? We define the efficiency of a package as the ratio of the area of the bottoms of the cans to the area of the bottom of the box. _____

   _____

2. Why is that definition reasonable?_____

   _____

3. What is the efficiency of the standard six-pack?

   _____

4. Express the efficiency as a percent. _____

5. Calculate the efficiency for a square-based one-pack. Is it the same as, or different from, the six-pack? Why did the answer turn out the way it did? _____

   _____

   _____

6. What would you expect the efficiency to be for a square-based four-pack? Calculate and check your prediction. Can you generalize to any pack that consists of a combination of square-based one-packs? _____

   _____

7. Calculate the efficiency of a standard, rectangular six-pack of cans if the radius of the cans is 4 cm, 5 cm, and 6 cm. Explain what you notice. _____

   _____

8. Use an algebraic argument to determine what happens in problem 7 by calculating the efficiency of the standard six-pack if the radius of the can is r cm. _____

   _____

You have been looking at rectangular boxes. What happens if you consider other shapes for the bottoms of the boxes? This activity is about parallelograms. Assume that the radius is *r* cm. Explain your calculations.

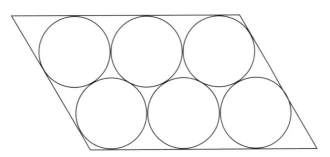

1. What is the area of the parallelogram? _____
   (*Hint:* The following figure shows many radii. Use it to show that the little triangle is equilateral, and draw some conclusions about the angles in the figure. Then use what you know about 30-60-90 triangles or the Pythagorean theorem to find the base and height of the parallelogram.)

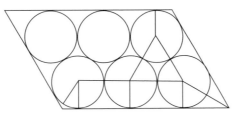

2. What is the efficiency of the parallelo-pack? _____

   _____

3. How does the efficiency of the parallelo-pack compare with that of the rectangular six-pack? _____

   _____

   _____

The rectangular six-pack arranged the cans in a stack position, whereas the parallelo-pack uses a staggered arrangement.

4. Discuss what effect you think the stack and the stagger methods might have on the efficiency of each shape of package. _____

   _____

5. In what other considerations about the parallelo-pack might manufacturers, wholesalers, distributors, retailers, and consumers be interested? _____

   _____

   _____

As a new employee of the Kula Kola Company, you have been waiting for an opportunity to demonstrate your abilities. Several teams have been selected to design a new packaging unit for the distribution of the various soft drinks produced by the company. You are expected to present to the company's executives a convincing argument that your design is the best in terms of efficiency. You know that the company's executives are most impressed when new ideas are supported by mathematical analysis rather than by approximate measurements.

It is rumored that a rival company, Popzi, has plans to unveil a different package in an upcoming advertising campaign. You have overheard some of the vice-presidents whispering, "Why would two rows of three cans be the best packaging arrangement for the six soft-drink cans? Should we consider a different number of cans as well as a new arrangement of cans?" Because the pressure is on from Popzi, your team has only a short time to complete your design and prepare a presentation.

You have been given twelve soft-drink cans to use. You can use any other tools your team has available. Be sure to submit the drawings of all your attempts as part of this investigation. You will need to keep track of the information and measurements for your different packages, since they might be used for further exploration.

Your goal: Make the most efficient package possible for sets of three cans or more. Remember that we define efficiency as the ratio of the area of the bottoms of the cans to the area of the bottom of the box.

1. Find the efficiency of each of the following containers:

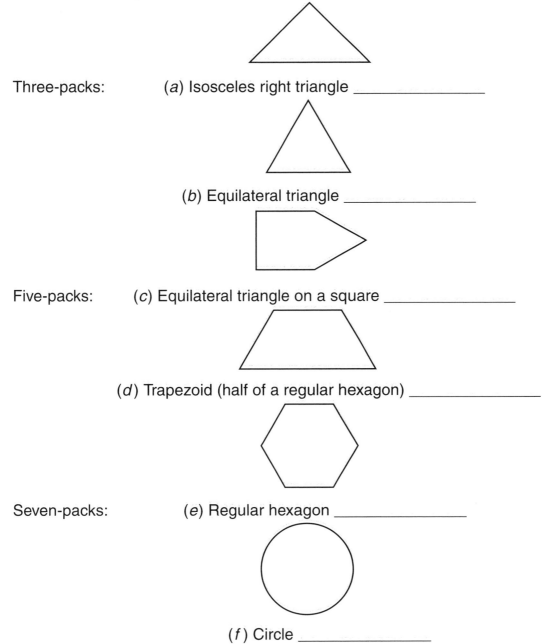

Three-packs: (*a*) Isosceles right triangle _____

(*b*) Equilateral triangle _____

Five-packs: (*c*) Equilateral triangle on a square _____

(*d*) Trapezoid (half of a regular hexagon) _____

Seven-packs: (*e*) Regular hexagon _____

(*f*) Circle _____

The design of soft-drink containers involves many other considerations besides efficiency as we have defined it. One particular factor to examine is how the package fits in its environment. Can the boxes be placed next to one another without wasting space? The process of filling a plane with a particular shape is called *tessellating the plane.*

2. Which of the foregoing shapes tessellate the plane? Trace each shape and make a model that you can use to show how well the shape fills the plane of a clean sheet of paper.

3. In your investigations, you have probably discovered that the rectangular six-pack is not the most efficient of all the shapes and that it is not the only shape that tessellates the plane. What other reasons might explain why the rectangle is the shape of choice?

From the *Mathematics Teacher*, March 1997

More Functions of a Toy Balloon

Loring Coes III

April 1997

Edited by Henri Picciotto, *The Urban School, San Francisco, California*

TEACHER'S GUIDE

Like all teachers, I am constantly looking for economical but memorable activities that illustrate important ideas in mathematics. Toy balloons, inexpensive and entertaining, offer some interesting opportunities to connect algebra and geometry. In the following activities, students can measure the speed of a released balloon and investigate the subtle relationship between the volume of a sphere and its surface area. In addition, the activities can prompt a rich and revealing discussion about why the experimental results are not the same as those we would expect theoretically.

Some related experiments were described in an earlier *Mathematics Teacher* article, "The Functions of a Toy Balloon" (Coes 1994). For that activity, students investigated—

- flight duration as a function of the number of breaths used to inflate the balloon,

- circumference of the balloon as a function of diameter, and

- diameter as a function of the number of breaths.

In those experiments, students found that flight duration is more or less proportional to the number of inflation breaths, circumference is proportional to diameter with something close to π as the ratio of proportion, and diameter is roughly proportional to the 1/3 power of volume. We used one breath as one unit of volume.

Several other questions come up as a result of the first activity:

- How far do the balloons really travel? Measuring actual distance is difficult because balloons almost always swerve and spiral in irregular paths.

- How fast do they go? Without knowing distance, we cannot calculate speed.

And a different kind of question also arises:

- How does surface area vary with diameter and volume?

The following activities look at those questions. You may wish to do the experiments from the earlier article first, and students may naturally pose some of the questions noted here. I have, however, done the activities independently, and you can choose how the materials best fit your program.

Grade level: The first experiment, sheets 1A and 1B, can be done with any high school students. Second-year-algebra or precalculus classes are good places for the investigation of surface area, sheets 2A and 2B.

Time: Sheets 1A and 1B are great for an extended time period. Ninety minutes should be enough to set up the experiment, collect data, clean up (sheet 1A), and discuss results (sheet 1B). Alternatively, each sheet can be done in one forty-five-minute period.

Sheets 2A and 2B can be done well in a single forty-five-minute period, although questions 6–9 may be good homework problems for the group, with some follow-up time the next day.

Materials: Sheets 1A and 1B: Several different sizes of balloons—large and small spherical ones, along with some oblong ones—about three balloons for each group of four students; stopwatches, one per team; monofilament fishing line in fifteen- to twenty-meter lengths (teams can share one line, if necessary); plastic drinking straw, cut in five-centimeter lengths, one per team; measuring tapes—one about ten meters long and several shorter ones for measuring the balloon's diameter; and pencil, paper, and recording sheets

Sheets 2A and 2B: Dark felt-tipped pens or markers; small transparent rulers for measuring squares on the balloons, such as flexible rulers made on transparency film; big bow calipers and large measuring tapes to measure the balloon's diameter; and a graphing calculator or a computer program like Cricket Graph III (CAI 1993)

Loring (Terry) Coes III is chair of the mathematics department of Rocky Hill School in East Greenwich, Rhode Island. He is interested in linking algebra and geometry.

Directions: Sheets 1A and 1B: The purpose of the fishing-line track is to ensure that the balloon travels in a straight line. You may precede the experiment by showing how a balloon that is let go without the use of the straw-and-fishing-line apparatus flies around randomly. To the extent possible, encourage the teams to set up their own tracks.

Although students may understand the concept of speed fairly well, they may not have had many opportunities to measure it. It will be worthwhile to talk about what a rate really is—a ratio showing how one quantity (here, distance) varies with another (here, time).

During the experiments, students may notice that the balloons do not have a constant speed. Talking about the concept of average speed may be fruitful and interesting. Watching the balloons change speed may prompt a good discussion of the difference between the average speed and the speed at any given moment.

Some incidental geometry. Students should experiment with different methods of taping the balloon to the straw. They will find that the *axis of thrust* should be parallel to the fishing line. If the thrust is not parallel, the balloon will wobble and lurch and will not go very far. Use this opportunity to show some physical models of *parallel, intersecting,* and *skewed* lines. (See **fig. 1.**) The distinctions really mean something in the experiment!

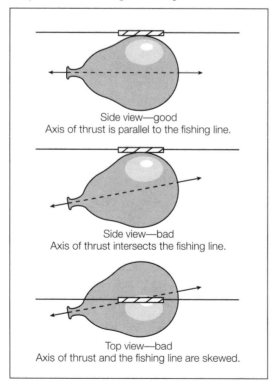

Side view—good
Axis of thrust is parallel to the fishing line.

Side view—bad
Axis of thrust intersects the fishing line.

Top view—bad
Axis of thrust and the fishing line are skewed.

Fig. 10.1. Models of parallel, intersecting, and skewed lines

Give students a chance to discover these principles. They will probably be able to say that the fishing line and the balloon axis must be parallel. If students do not reach this conclusion themselves, the experiment will at least afford a good opportunity for discussion.

Practice makes perfect. If students have an opportunity to repeat the experiment, you will find that they get better at measurement, experiment design, and data collection. Using a stopwatch, for example, can be tricky, and it is worth practicing. I have also noticed that many otherwise competent mathematics students have trouble using a tape measure! Here is a good opportunity to practice.

Wind. The balloons will be highly sensitive to wind. A calm day is good, and a long, closed hallway or gymnasium is even better.

Unit conversion. If your students have trouble with question 8 on sheet 1B, you may introduce this format for the calculation, explaining that each fraction except the first and last equals 1 and that the units "cancel":

$$\frac{1 \text{ meter}}{1 \text{ second}} \bullet \frac{3.28 \text{ feet}}{1 \text{ meter}} \bullet \frac{1 \text{ mile}}{5280 \text{ feet}} \bullet \frac{60 \text{ seconds}}{1 \text{ hour}}$$

$$\bullet \frac{60 \text{ minutes}}{1 \text{hour}} = \frac{?? \text{ miles}}{\text{hour}}$$

Sheets 2A and 2B: In this experiment, students investigate how small squares of surface area grow with the number of breaths and with the diameter of the balloon. Students will find, though, that balloon material does not expand uniformly over the entire surface, so it is a good idea to *sample* the surface expansion by measuring a few of the squares over the balloon surface. Students then average the three samples.

Regression. Many graphing calculators and computer programs allow *regressions,* where the program fits the data entered to a function form requested. Typically, one variable is put in one column, and the other variable, in another column. The program then performs a linear, polynomial, power, or exponential regression. The program generates the parameters for the particular function form requested. In these activities, students do a *power regression* of the form $y = ax^b$. Students enter x-data and y-data as number pairs, and the program automatically returns the values of a and b that best fit the data. The manuals for regression programs and calculators provide more information.

Using Activities from the *Mathematics Teacher* to Support *Principles and Standards*

A power regression assumes that for $x = 0$, $y = 0$. This assumption is appropriate here, since either zero diameter or zero volume implies zero surface area.

We should expect the surface areas to grow in proportion with the square of the diameter because the surface area of a sphere is given by the formula $A = 4\pi r^2$ and the diameter is just twice the radius.

Relating the number of breaths to surface area is more complex. If we assume—cautiously!—that one breath is one unit of volume, then we can reason in the following way:

- The surface area of a spherical balloon is $4\pi r^2$.
- The volume is $(4/3)\pi r^3$, so the surface area should be proportional to the 2/3 power of the volume.

A way to express surface area as a function of volume follows:

- We need to get $4\pi r^2$, which is approximately $12.57r^2$, from the expression $(4/3)\pi r^3$, which is approximately $4.19r^3$.
- Take the 2/3 power of the volume expression:

$$V^{2/3} \approx (4.19r^3)^{2/3} \approx 2.60r^2$$

- To get $12.57r^2$, multiply $2.60r^2$ by approximately 4.83, so the surface area $\approx 4.83 \cdot 2.60r^2 \approx 4.83V^{2/3}$.

Since we are measuring only small sections of the surface area, we hope to see only a proportional relationship, like this one:

Surface area = constant $\cdot$ breaths$^{2/3}$

Of course, balloons are not exactly spheres, but the argument can be made in a more general way by thinking about dimension, in the following manner:

- Surface area, being two-dimensional, is proportional to the square of linear dimensions: $A = kd^2$.
- Volume, being three-dimensional, is proportional to the cube of linear dimensions: $V = md^3$.
- The equation $V = md^3$ gives

$$d = \sqrt[3]{\frac{V}{m}}$$

so

$$A = k\left(\sqrt[3]{\frac{V}{m}}\right)^2 = \frac{k}{m^{2/3}}V^{2/3} = cV^{2/3},$$

where c is a constant.

This argument works for any three-dimensional object that increases in size while remaining similar to itself. To the extent that similarity is not exactly preserved, the results obtained in the regression will differ from the ones in the theoretical argument, even if the measurements were exactly accurate.

Conclusion: The theoretically expected exponents in the regressions should be 2 for the surface area as a function of diameter and 2/3 for the surface area as a function of volume. Be sure to discuss what might cause the results of the regression to differ from the expected. Aside from the foregoing point about whether the shape remains similar to itself, a number of reasons might explain why groups may not get expected results. Caution against trying to blame a member of the team. Here are two possibilities; you or your students may think of others: (1) the balloons themselves may expand in an irregular way or (2) it is difficult to measure breaths to make them consistent.

Assessment: Before the activities, let the students know how they will be assessed. I suggest assigning from 0 through 4 points for each subsequent question, with 4 for exemplary work and 0 for no response. This assessment will be easier to carry out if you ask for written laboratory reports from individuals or groups.

Sheets 1A and 1B:

- How well do students set up the track and understand the best positioning of the balloon?
- Are measurements done well, and do students get better at measurement as the experiment moves along?
- How well do students understand that in computing *average* velocity, we are *not* measuring speed at a particular moment and we are not measuring changes in speed?

Sheets 2A and 2B:

- How well do students understand the relationship between volume and surface area?
- How well do students understand that the surface area can be expressed as a *function* of volume?

It is generally not a good idea to evaluate students' work on how well their data match theoretical expectations, since that approach can encourage some to "cook" their data to get closer to the expected results.

Since this activity is a team effort, especially for sheets 1A and 1B, I suggest assigning from 0 through 4 points for the quality of teamwork shown by each group. In assigning that grade, think about these questions: Did they work well together? Were jobs distributed well? Did students support one another?

Selected answers:

Sheet 1A: The actual data will vary. Typically, a balloon can travel ten meters in three seconds.

Sheet 1B: See also the Teacher's Guide.

3–5. Answers will vary. Distance will generally be proportional to breaths. Surprisingly, speed often decreases with the number of breaths because the highly inflated balloons offer so much wind resistance.

6–7. Balloons often speed up when they get small because the small size gives less wind resistance.

8. 2.24 MPH

Sheet 2A: Data will vary.

Sheet 2B: See the Teacher's Guide.

References
Coes, Loring, III. "Activities: The Functions of a Toy Balloon." *Mathematics Teacher* 87 (November 1994): 619–22, 628–29.
Computer Associates International (CAI). Cricket Graph III. Islandia, N.Y.: CAI, 1993. Software.

In this experiment, you will inflate a balloon, attach it to a track made of fishing line, and then measure the distance and time that it travels. You will inflate the balloon with different numbers of breaths and analyze how these numbers affect distance, time, and speed.

Assign roles to each member of the group: an inflater to blow up the balloon, two people to tape the balloon to the straw, a timer, a distance measurer, and a data recorder. Tie one end of the fishing line to a door latch or some other secure spot. The setup will look like the following diagram:

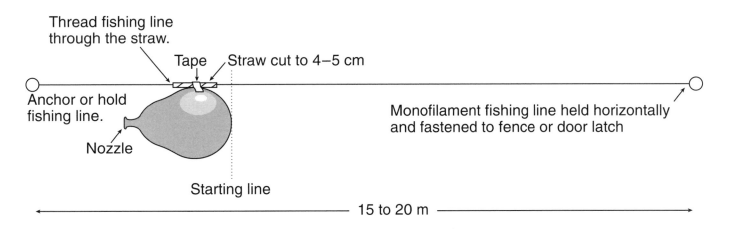

Use chalk or tape to mark distances from a starting line at five-meter intervals, which will make it easier to measure the distance that the balloon travels.

- Inflate a balloon with one breath. Tape it carefully to the straw.

- Release the balloon and measure both the flight time and the distance.

- Repeat with two breaths and so on to complete the table.

Describe the balloon. _____

No. of Breaths	Distance (m)	Time (s)	Speed (m/s)
1			
2			
3			
4			
5			
6			

If time allows, repeat the experiment with a balloon of a different shape to create another similar table.

From the *Mathematics Teacher*, April 1997

1. Think about the axis of *thrust* in the balloon—a line going through the nozzle opening to the very front of the balloon. What relationship should this axis have with the fishing line to maximize the flight?

2. Make graphs using the data from the table.

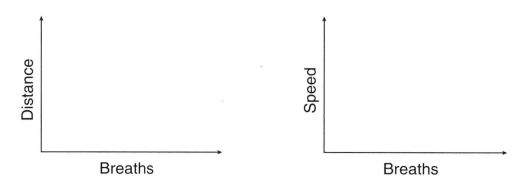

3. Describe the graphs. What patterns do you see?

4. How does the number of breaths affect the distance traveled?

5. How does the number of breaths affect the speed?

6. What do you think causes a balloon to be fast or slow?

7. You have computed the *average speed* for the balloons during this experiment. Do the balloons travel at the average speed all the time? Explain.

For many people, miles per hour is the most common way to measure speed. It is difficult to compare meters per second (m/s) and miles per hour (mi/hr).

8. Convert 1 meter per second to miles per hour:

1 m/s = _____ mi/hr

Here are the necessary conversion factors:

1 meter ≈ 3.28 feet
1 mile = 5280 feet
1 hour = 60 minutes
1 minute = 60 seconds

9. Find the average speed of your balloon in miles per hour.

In this experiment, you will inflate a balloon with one breath, then draw small (1-centimeter) squares at three locations on the balloon. You will measure the growth of the squares as you inflate the balloon further.

Assign roles to each member of the group: an inflater; a caliper handler; a diameter measurer, who will use a large tape measure; a square measurer, who will use a small transparent ruler; and a recorder.

To measure the diameter consistently, mark two spots on opposites sides of the uninflated balloon.

When the balloon is inflated with one breath, draw three 1-centimeter squares on the balloon in different places and label the squares *A*, *B*, and *C*.

When you measure the area of the square, be sure to measure both length and width—the two dimensions may not be expanding at the same rate!

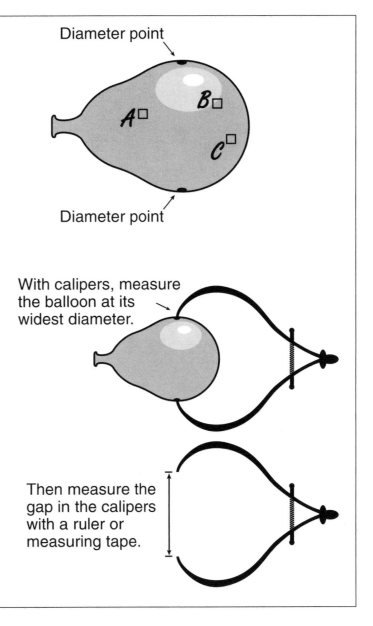

Diameter point

Diameter point

With calipers, measure the balloon at its widest diameter.

Then measure the gap in the calipers with a ruler or measuring tape.

Number of Breaths	Diameter (cm)	Square A		Square B		Square C		Average Area (cm²)
		Dimensions (cm × cm)	Area (cm²)	Dimensions (cm × cm)	Area (cm²)	Dimensions (cm × cm)	Area (cm²)	
1		1 × 1	1	1 × 1	1	1 × 1	1	1
2								
3								
4								
5								
6								

From the *Mathematics Teacher*, April 1997

1. Make two graphs from the data in the table.

   (a) Put the diameter on the x-axis and the average area of the squares on the y-axis.

   (b) Put the number of breaths on the x-axis and the average area of the squares on the y-axis.

2. Describe each graph.

3. Use the table to think of "Average Area" as a function of "Diameter." If possible, use a graphing calculator or a computer program to do a power regression on the data representing these two quantities. Your graph in question 1(a) is a graph of these data.

4. Think about the following formulas for a sphere, where r is the radius:

   • Diameter = 2r

   • Surface area of an entire sphere = 4πr2

   Use these formulas to explain why it is reasonable for the area-diameter graph, from question 1(a), to be quadratic. How does this result compare with your result from question 3?

5. Use the table to think of "Average Area" as a function of "Number of Breaths," and do a power regression on the data representing these two quantities. Your graph in question 1(b) is a graph of these data. What do you get?

6. Discuss this problem with your team. Recall that the volume of a sphere is given by the formula

$$V = \frac{4}{3}\pi r^3 \ .$$

   Use this fact, and the formula in question 4 for the surface area of the sphere, to explain why—theoretically, anyway!—you might expect your regression equation in question 5 to show "Average Area" as approximately proportional to the 2/3 power of "Number of Breaths." Remember that the average area of the squares is roughly proportional to the total surface area of the sphere. It is also reasonable to assume that the volume of a balloon is roughly proportional to the number of breaths.

7. Use algebra to calculate the missing coefficient on the basis of the formulas given earlier:

   Total surface area of a sphere = _____ $V^{2/3}$

8. Your answer to question 7 gives the total surface area as a function of volume. Since you are sampling only small portions of the total surface area, how would you expect your regression equation in question 5 to be different from your answer to question 7?

9. Explain why your regression equation in question 5 might not be what you expect.

Activities for the Logistic Growth Model
or, Invasion of the Killer Moths

Robert Iovinelli

October 1997

Edited by Loring (Terry) Coes III, Rocky Hill School, East Greenwich, Rhode Island

TEACHER'S GUIDE

"Why has this little bug not taken over the world ?"

When students begin to study exponential growth and they see a model for the rapid growth of, say, an insect population, they may wonder, "Why has this little bug not taken over the world if it can grow so fast? It is a reasonable question that allows the introduction of a function that models the situation better than the straightforward exponential function. The growth of the population of a species, when first introduced into an environment, can be subdivided into several different stages only one of which is exponential.

The initial-growth stage is a small increase in numbers. Then, once the specimen adapts to the environmental conditions, its population grows very rapidly. This growth produces a doubling type of effect that is seen in terms of a graph as exponential growth. The next stage is a dampening of the growth due to limiting factors, such as competition for food and safe shelter; predation on the species; parasites that can weaken their host and produce death by disease; and overcrowding, which can sometimes cause members of the species to eliminate their own kind. These curtailments in a population are known as density-dependent limiting factors. Additionally, the population will have to endure density-independent factors of natural cycles in the environment, such as droughts, floods, and hurricanes. Because of these conditions, the population of the specimen reaches a steady state in size known as the *carrying capacity* of the species in a closed environment. This situation does not mean that the size of the population declines. Rather, the average birth and death rate become equal.

The following population model is exponential at an early stage, then shows the dampening effects of environmental restrictions on a large population growth. Figure 1 shows a graphical interpretation of these concepts. The graph is interpreted by biologists as having four distinct stages, or intervals:

(1) The initial-growth stage
(2) The exponential-growth stage
(3) The dampened-growth stage
(4) The equilibrium-growth stage

Consider as an illustration of this model the introduction of a flu virus into a closed environment by means of a single infected individual. As the diseased person comes into contact with other people, these infected people start to slowly spread the virus, and we can measure the spread of the virus by the total number of infected individuals. For a while, the infection rate grows swiftly because during the beginning stages, very few people of the total population have yet been infected. The high ratio of people who do not have the disease to people who do have the disease produces the stage of rapid exponential growth.

As more and more people become sick, the number of possible contacts with new, uninfected hosts decreases. This density-dependent factor reduces the rate of infection to less than exponential. As the ratio of noninfected to infected persons become smaller, the chances of a contact between noninfected and infected individuals becomes smaller.

The number of infected people, although still growing, is growing at an ever-decreasing rate. Finally, the virus has almost no one left to infect,

Editorial Comment: The author used a spreadsheet to complete Sheet 3, although he suggests it could be rewritten to use a graphing calculator. My students prefer using the LIST and TABLE features of their graphing calculators. A related article, "Teaching the Logistic Function in High School." appeared in April 2002. It contains a program for a graphing calculator to model the spread of an infection.

Bob Iovinelli, is the technology coordinator for Grapevine-Colleyville Independent School District, Grapevine, Texas 76051-3897, and teaches at Tarrant County Junior College, Fort Worth, Texas 76102. His interests are in linking mathematics and science through the use of technology.

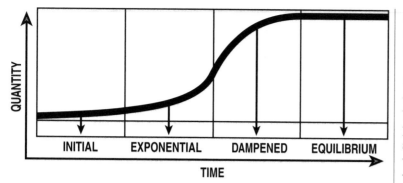

Fig. 1. Intervals showing four stages of growth

producing a steady-state condition. A representation of this scenario is given by an equation of the form

$$y = \frac{a}{1 = + be^{-cx}}$$

The graph of an equation of this type is called a *logistic curve*.

Haeussler and Paul (1993) explain that this equation can be derived by solving an appropriate differential equation, which would show that "the rate of growth is proportional to the product of the population size and the difference between the maximum size and the population size" (p. 906).

The graphical representation is a serpentine-shaped curve that starts out with a slow increase; changes into an exponential-growth curve, as seen by its upward concavity; then starts to slow down, noted by its downward concavity, as it approaches the carrying capacity of the environment. The carrying capacity is represented by the limit of this function. The point at which the concavity changes is called a *point of inflection* (see fig. 2).

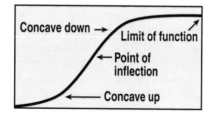

Fig. 2. A logistic curve

A total of eight people are infected.

An in-class simulation of this growth process can be modeled with the aid of a calculator to generate random numbers. In general, a random-number generator will produce a number between 0 and 1. For the simulation that my class did, we imagined a total population of 100 individuals. Each number from 0 through 99 represented an individual, with the label 0 used to portray the original host. You may want to label each individual

with a person's initials to reduce the confusion of using a number both to label a person and to show the quantity of infected people.

In my classroom, random numbers were generated by a TI-82 graphing calculator using the expression rand*100. The seed value was set to zero to produce integral values, so that the procedure produced equally likely outputs from 0 to 99 inclusive. In the classroom, you will need to experiment with the particular calculator used to determine which method works best with your equipment, since the actual procedure will depend on the type of calculator used.

The moderator can write the list of numbers from 0 through 99, representing individual people, on a side board in the classroom. A calculator attached to a view screen will allow the students to see the numbers as they are randomly generated and to record the outcome of each "day's" event in a data table. The numbers are crossed out as the given individual becomes infected, so we begin with 0 crossed out. For the first day's event, the original host infects a person represented by a randomly generated number, say, 42. Crossing out 42 on the board list means that two people now have the virus: 0 and 42.

$\cancel{0}, 1, 2, \ldots, 40, 41, \cancel{42}, 43, \ldots, 98, 99$

The number of people infected with the flu virus for the second day's event is determined by generating two more numbers, accomplished by pressing the **ENTER** key twice. As an illustration, assume that person 0 infects person 3 and person 42 infects person 14. The total number of people with the flu is now four. The list will appear as follows:

$\cancel{0}, 1, 2, \cancel{3}, 4, 5, \ldots, 13, \cancel{14}, 15, \ldots,$
$41, \cancel{42}, 43, \ldots, 98, 99$

Continue in this manner for each day's event. For day 3's event, suppose the following:

0	infects	5
3	infects	13
14	infects	21
42	infects	77

Now the people represented by 0, 3, 5, 13, 14, 21, 42, and 77 all have the flu virus and are crossed out on the list; that is, a total of eight people are infected.

The simulation will soon start to produce duplications. For example, the people represented by the numerals 5 and 14 might both infect the same person, for example, the person represented by the numeral 60. Also, a redundancy of contacts may occur. For example, person 14 might infect the already infected person 21. Another possibility is for an infected person to generate his or her own number. These duplications and repetitions are a desired aspect of the simulation, as they cause the exponential-growth stage to change into the dampened-growth stage. As the number of infected individuals grows, the simulation of each day's events becomes more time-consuming, so the teacher may not want to continue the activity until 100 percent of the population is actually infected with the virus.

A graph of the results, showing number of days (x) and number of people infected with the flu virus (y), will produce a good approximating visual model of the logistic curve.

Our classroom experiment produced the data in table 1. These data can be approximated by the logistic equation

$$y = \frac{100}{1 + 99e^{-0.7x}}.$$

TABLE 1		
Data From Classroom Experiment		
Daily Events	Total Number Infected	Daily Increase
Start	1	0
1	2	1
2	4	2
3	7	3
4	14	7
5	24	10
6	39	15
7	56	17
8	72	16
9	84	12
10	91	7
11	95	4
12	97	2
13	98	1
14	99	1

The motivation for pursuing logistic growth models is to answer the question "Why has this bug not taken over the world?" The simulation presented previously is the lead-in material for the activities. After the class exploration on viruses is presented, the teacher can tell the students that this type of growth pattern can be modeled by a certain type of equation, one that has the general form

$$y = \frac{a}{1 + be^{-cx}}.$$

Although the teacher will not want to explain to students the concepts of calculus used to derive this equation, the teacher can state that this model foreshadows ideas that will be learned in calculus. Next, the teacher can introduce the terminology for the four intervals encountered in a logistic growth curve, the nature of the curve's shape, and the reasons why the curve progresses through the different stages. The activity sheets are then sequentially used.

In my classroom, the initial presentation of modeling logistic curves was done after students had been introduced to the transcendental number e. For classes that have not covered this topic, introducing e as a constant similar to π will allow the use of the activities as written. For teachers who are not comfortable with this approach, the worksheets can be rewritten to use the equation

$$y = \frac{a}{1 + b * 2^{-cx}}.$$

Changing the base from approximately 2.7 to 2 still produces the desired serpentine nature of the curve.

The process for a teacher to follow depends on the needs and desires of her or his particular situation. The work presented in this article is a guideline that contains the necessary ingredients for teachers to introduce a model that is not usually encountered in precalculus courses.

Sheet 1: This worksheet reinforces the concepts presented in the class simulation with the x-axis representing "Number of Days" and the y-axis representing "Number of People Infected." The students need to approximately identify the interval $0 < x < 4$ as the initial-growth stage, the interval $4 < x < 8$ as the exponential-growth stage, the interval $8 < x < 16$ as the dampened-growth stage, and the interval $x > 16$ as the time when the camp is completely infected. The total number of people is 300, and the graph ceases to show exponential growth in the neighborhood of the eighth day.

Sheet 2: This worksheet uses the "rumor" equation

$$y = \frac{150}{1 + 149e^{-1.5x}}.$$

The point of inflection occurs in the neighborhood of the point (3.34, 75). The objective is for the student to construct a straight line within an interval close to the point of inflection. The difference between the y_1-values of the straight line and the y-values of the "rumor" curve will have a change in algebraic sign. This change shows that the line is below the curve at one place and above it at another. This fact will lead to the validation of a conjecture stating that the logistic curve is changing from an exponential phase to a dampening stage. The equation $y_1 = 56x - 112$, derived from the ordered pairs $(3, 56)$ and $(3.5, 84)$, was the most frequent choice of the students.

Table 2 shows the value table, and figure 3 shows the straight line's position in relationship to the rumor curve. The students' actual answers will depend on the points chosen for the line and hence will vary.

TABLE 2							
The Value Table for the Rumor Equation							
Time in Hours	2	2.5	3	3.5	4	4.5	5
Rumor curve's values	17	33	56	84	109	128	138
Straight line's values	0	28	56	84	112	140	168
Difference	17	5	0	0	-3	-12	-30

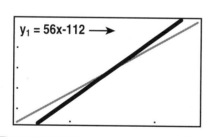

$y_1 = 56x{-}112 \longrightarrow$

Fig. 3. The straight line's position in relation to the rumor curve

Sheet 3: A computer and spreadsheet are used for this activity. If this type of technology is not available, then activity 3 can be rewritten to use, for example, a graphing calculator, or, as another option, the computations can be done with an x-y-value table. As written, one appropriate set of algorithms for using a ClarisWorks or Microsoft Works spreadsheet is seen in table 3.

Using the spreadsheet will produce 224, 431, 718, 954, 946, 702, 417, 216, 104, 48, 22, 10, 4, 2, 1 in the difference column. The number 954 will be the desired point in the neighborhood at which the curvature of the graph changes from concave-up to concave-down and, hence, shows the change from exponential growth. The results are seen in table 4.

TABLE 3			
Spreadsheet Algorithms			
A	B	C	D
1. Days	Computation	Number of fish	Difference
2. 0	=1		
3. =1	=Exp(-0.8*(A3))	=5000/(1+24*(B2))	
4. =A3+1	Fill Down	Fill Down	C4-C3
5. Fill Down	---------	---------	Fill Down

TABLE 4			
Spreadsheet Results			
A	B	C	D
Days	Computation	Number of Fish	Difference
0	1.00e + 0		
1	4.49e − 1	200	
2	2.02e − 1	424	224
3	9.07e − 2	855	431
4	4.08e − 2	1574	718
5	1.83e − 2	2527	954
6	8.23e − 3	3473	946
7	3.70e − 3	4175	702
8	1.66e − 3	4592	417
9	7.47e − 4	4808	216
10	3.35e − 4	4912	104
11	1.51e − 4	4960	48
12	6.77e − 5	4982	22
13	3.04e − 5	4992	10
14	1.37e − 5	4996	4
15	6.14e − 6	4998	2
16	2.76e − 6	4999	1
17	1.24e − 6	5000	

The what-if questions allow the students to explore the nature of a logistic curve by changing the parameters a and b in the general equation

$$y = \frac{a}{1 + be^{-cx}}.$$

Question 2 will produce a graph and chart that show the original number of fish stocked to be 100 and the total capacity to be 5000 (fig. 4).

Question 3 will produce a graph and chart that show the original number of fish stocked to be 40 and the total capacity to be 2000 (fig. 5).

Using Activities from the *Mathematics Teacher* to Support *Principles and Standards*

Day(s)	Number of Fish
0	
1	100
2	217.229617116
3	459.013349373
4	918.243339101
5	1668.1413064
6	2635.09290541
7	3563.1376454
8	4233.00017061
9	4623.56630881
10	4823.54179533
11	4919.14079302
12	4963.34115592
13	4983.46134652
14	4992.55514337
15	4996.6520652
16	4998.49512104
17	4999.32370221

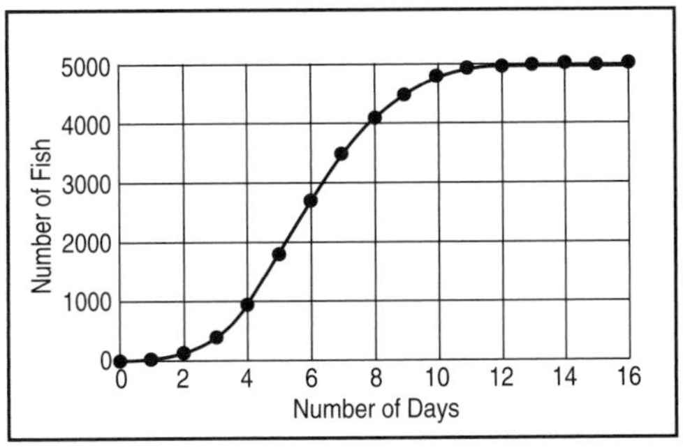

Fig. 4. Population growth for question 2

Day(s)	Number of Fish
0	
1	40
2	86.8918468464
3	183.605339749
4	367.29733564
5	667.256522558
6	1054.03716216
7	1425.25505816
8	1693.20006824
9	1849.42652352
10	1929.41671813
11	1967.65631721
12	1985.33646237
13	1993.38453861
14	1997.02205735
15	1998.66082608
16	1999.39804842
17	1999.72948088

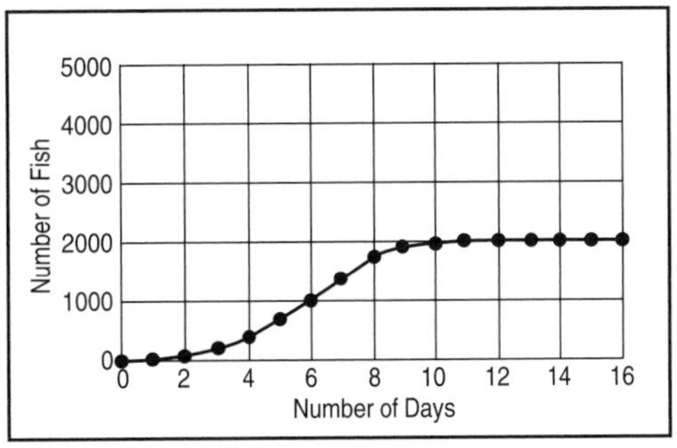

Fig. 5. Population growth for question 3

The answers to question 4 will vary according to the choices made by the students. The main environmental factor that limits the population's growth, for question 5, is competition for food.

Sheet 4: The equation used is

$$y = \frac{10\,000}{1 + 49e^{-1.2x}}.$$

The parameters of a and b, respectively, are 10 000 and 49. The initial value of 200 is arrived at when $x = 0$, using $e^0 = 1$, because

$$y = \frac{10\,000}{1 + 49(1)} = 200.$$

Also, as x grows without bound, then $e^{-1.2x}$ approaches 0, so y approaches

$$y = \frac{10\,000}{1 + 49(0)},$$

which equals 10 000. The graph (fig. 6) shows the point of inflection to be in the neighborhood of the point (3.243, 5 000). Hence, the population ceases to grow exponentially after approximately three months and one week.

Question 4 explores the rate of growth by changing c, the coefficient of the exponent. Although the main objective is for the student to verify or deny the class's conjecture, it is important for the student to notice that neither the initial nor the maximum values are different for the two equations (see fig. 7).

Question 5 is an opportunity for the student to show mastery of the subject. It presents the in- structor with the latitude to assign the question

The population ceases to grow exponentially

Activities for the Logistic Growth Model

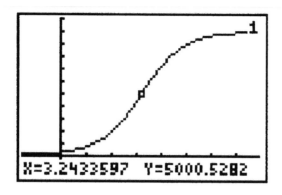

Fig. 6. The point of inflection is in the neighborhood of (3.243.5 000).

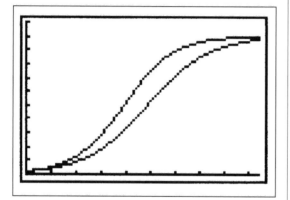

BIBLIOGRAPHY

Dugopolski, Mark. *College Algebra*. Reading, Mass.: Addison-Wesley Publishing Co., 1995.

Haeussler, Ernest F., Jr., and Richard S. Paul. *Introductory Mathematical Analysis for Business, Economics, and the Life Sciences*. 7th ed. Englewood Cliffs, N. J.: Prentice Hall, 1993.

Larson, Roland E., Robert P. Hostetler, and Bruce H. Edwards. *College Algebra, a Graphing Approach*. Lexington, Mass.: D. C. Heath & Co., 1993.

———. *Calculus*. 5th ed. Lexington, Mass.: D. C. Heath & Co., 1994.

Tan, S. T. *Mathematics for the Managerial, Life, and Social Sciences*. Boston: PWS Publishing Co., 1996.

The author wishes to express a special thanks to the editors and reviewers who offered detailed, accurate, and excellent suggestions during the rewriting stages of this article.

as a portfolio extension by having the student do research on insects, besides the gypsy moth, that are destructive to crops, such as the browntail and tussock moths. In addition to the book references in a library, such CD-ROM encyclopedias as *Encarta, Comptons,* and *Groliers* are excellent resources. Institutions or individual students who are fortunate enough to have access to the Internet can use Microsoft School House or Ask ERIC to do research.

Using Activities from the *Mathematics Teacher* to Support *Principles and Standards*

In a small summer camp, one student enrolled with a contagious flu virus. The spread of the virus through the camp's population was displayed by the following graph.

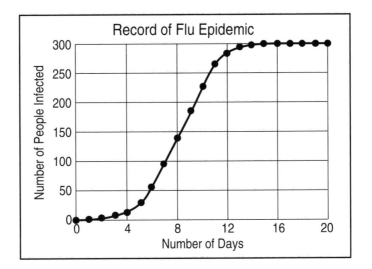

1. Write a short essay describing the overall shape of the graph. Include in your description the intervals on the horizontal axis where the graph—

    a) shows the virus beginning to spread through the population of the camp,

    b) shows the greatest increase in the rate of infection of the camp's population,

    c) shows where the rate of infection seems to be slowing down,

    d) shows where the population of the camp is completely infected.

    What do you think happened that caused the graph to change shape from one interval to the next? _____

    _____

    _____

    _____

2. What do you think is the total number of people enrolled in the camp? _____

3. On what day does the rate of infection appear to start to slow down? _____

From the *Mathematics Teacher*, October 1997

Last spring, 150 students attended an all-night prom dance. One student started spreading a rumor that her best friends had broken up. The number of students that knew about the rumor in x hours is given in the following chart.

x: Time in hours:	0	0.5	1.0	1.5	2.0	2.5	3.0	3.5	4.0	4.5	5.0
y: Number of students:	1	2	4	9	17	33	56	84	109	128	138
x: Time in hours	5.5	6.0	6.5	7.0	7.5	8.0	8.5	9.0			
y: Number of students	44	147	148	149	149	150	150	150			

1. Make a graph of the given data by plotting points and connecting them with a smooth curve.

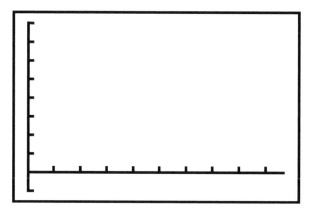

2. Identify the interval where the curve most resembles a nonhorizontal line. By choosing two points within this interval, derive the equation of a straight line.

   Your points: _____ Your equation: y_1 = _____

3. Fill in the chart below from the given information about the rumor curve (x, y) and the values computed from your equation (x, y_1). Find the difference between y and y_1, and make a conjecture as to what is happening to the rumor curve within the interval of your straight line.

Time in hours (x)										
Rumor curve's values (y)										
Straight line's values (y_1)										
Difference ($y - y_1$)										

4. What other interval contains a section of the rumor curve that looks like a straight line? Explain what is happening in this interval.

An ichthyologist decided that by introducing a new predator species of fish, for example, striped bass, into a small lake, their population would increase according to the logistic equation

$$y = \frac{5000}{1+24e^{-0.8x}},$$

where x is the number of days and y is the total population.

1. a) Make a table by setting up a spreadsheet that shows the number of days, the number of fish, and the daily increase in the number of fish. Make sure to calculate enough days so that all pertinent data are shown.

   b) Use the spreadsheet's charting capabilities to portray the data graphically.

   c) Identify approximately which point represents the day when the logistic curve changes from the exponential-growth stage to the dampened-growth stage.

2. a) Estimate the total number of fish and the original number of fish stocked if the equation was

$$y = \frac{5000}{1+49e^{-0.8x}}.$$

   b) Test your estimate by changing the spreadsheet to accommodate the change in the denominator from $1 + 24e^{-0.8x}$ to $1 + 49e^{-0.8x}$. Make a graph.

3. a) Estimate the total number of fish and the original number of fish stocked if the equation was

$$y = \frac{2000}{1+49e^{-0.8x}}.$$

   b) Test your estimate by changing the spreadsheet to accommodate the change in the numerator from 5000 to 2000. Make a graph.

4. Describe how the changes in the numerator and denominator changed the shape of the logistic graph. Make a generalized statement of how the numerator and denominator determine the total number of fish and the original number of fish stocked. Support your statement by showing examples different from those used on this worksheet. Using the charting capabilities of the spreadsheet to support your conclusion visually with a graph.

_____

_____

_____

_____

5. Describe what environmental factors might limit the bass population from continuing their exponential-growth rate. _____

_____

_____

_____

To predict the amount of defoliation that would be caused by the gypsy moth, a biologist counted 200 egg masses in a confined area. She predicted, on the basis of past data, that the carrying capacity of the given environment would produce 10000 killer moths if the logistic growth rate was $e^{-1.2x}$.

1. Derive an accurate mathematical model for this scenario using an equation of the form

$$y = \frac{(\quad)}{1 + (\quad)\, e^{-1.2x}}$$

where x is the time in months and y is the total number of egg masses.

2. Using a graphing utility, graph the equation. Copy the graph onto the coordinate box.

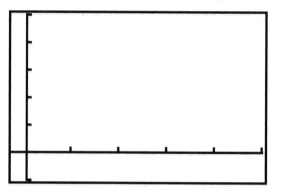

3. Use the ▐TRACE▐ and ▐ZOOM▐ functions to locate an approximate month and week when the growth of the population ceased being exponential.

4. Given the same scenario, what do you think would happen if the rate of growth was first changed to $e^{-1.0x}$, then changed to $e^{-0.8x}$? Verify or deny your conjecture by comparing both graphs. _____

_____

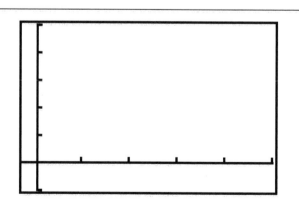

5. Write a story problem that depicts the destruction to crops by an insect. Present the data in your problem as sets of ordered pairs that resemble a logistic curve, then explain how to derive a logistic equation from the data. Include how you determined the values for the numerator and denominator. Explain how by trial and error you can approximate the rate of growth that is expressed as the coefficient of the exponent. Verify your results with a graph and value table. _____

_____

The Secret of Anamorphic Art

Art Johnson and Joan D. Martin

January 1998

Edited by Claudia Carter, *Mississippi School for Mathematics and Science, Columbus, Mississippi*

TEACHER'S GUIDE

It is 1655 and you are living in England, a supporter of the deposed royal family. Such a political persuasion might mean your life if it were known to the antiroyalist authorities. Nevertheless, you carry a portrait of the executed king, Charles I, with you at all times. Although everyone knows what Charles I looked like, you are confident that you will not be arrested. Why not? Simple—since this portrait was done in an anamorphic-art style, it cannot be recognized by anyone who does not know the secret of anamorphic art.

Anamorphic art refers to artwork that is indistinct when viewed from a normal perspective but becomes recognizable when the image is viewed from a different perspective or reflection. The term *anamorphic* is derived from two Greek words meaning *to change again*. Leonardo da Vinci was an early experimenter with anamorphic art. He produced black-and-white sketches that were distorted from a normal view but formed human faces when viewed from the extreme edge of the canvas.

> *Anamorphic art is indistinct when viewed from a normal perspective*

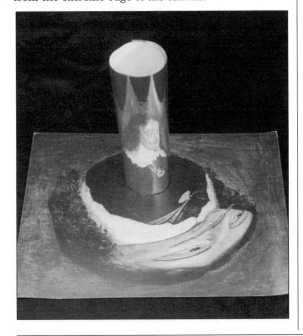

The anamorphic-art activity in this article uses the reflection of mirrored cylinders. Anamorphic art is thought to have originated in China about 700 years ago. The earliest known examples of anamorphic art using a mirrored cylinder are three Chinese specimens that date from about 1575. In 1630, the secret to making anamorphic drawings was revealed in *Perspective Cylinderique et Conique* by Vaulezard (Leeman 1976). Within a century, anamorphic art became all the rage, with many examples flooding the houses of the well-to-do throughout Europe. For the next 200 years, anamorphic art became as commonplace as wall murals are today. The invention of the daguerreotype changed all that. By 1860, the public's fascination with anamorphic art died out; instead, people focused on the photograph as a better way of reflecting the world around them.

Two main techniques are available for creating anamorphic art. The Chinese artists apparently drew their pictures while concentrating on the reflection in a polished cylinder or sphere. By contrast, European artists used the grid system presented in this activity.

Grade levels: This activity is suitable for either an algebra class or a geometry class. The activity requires some minimal skills from coordinate geometry and spatial visualization. If students have not used the terms reflect and translate, some intuitive introduction of the concepts will be needed.

Materials: Tape, reflective Mylar, a cardboard tube, and drawing materials, such as markers and colored pencils. The mirrored cylinders needed for

Art Johnson teaches at Nashua Senior High School, Nashua, New Hampshire. His interests are mathematics history and geometry, as well as research relating to proportional reasoning. Joan Martin is a mathematics/science specialist in the Newton Public Schools, Newton, Massachusetts, and works at the Mathematics Institute at Boston College. Her interests include the integration of mathematics with other disciplines.

Editorial Comment: The circles on sheets 4 and 5 have a smaller diameter than the tubes suggested by the author. In my class, a 100-mL graduated cylinder had the needed diameter. The activities could also be enlarged to accommodate a larger cylinder.

the activity can be made by taping the reflective Mylar or reflective origami paper around the outside of a cardboard tube, such as a paper-towel roll or a bath-tissue roll.

Procedure: Students should work in small groups so that they can help one another. It is also helpful for the teacher to circulate while students are working to ensure that they are all moving from a standard coordinate grid to an anamorphic one.

The activity sheets are organized to bring students from a traditional coordinate grid to an anamorphic grid, which is then used to produce anamorphic images in a mirrored cylinder. The first four activity sheets can be completed in a standard class period; the fifth sheet is an out-of-class assignment.

Sheet 1: In this activity, students are asked to translate triangle *ABC* on the coordinate plane by reflecting the vertices across the *x*-axis and the *y*-axis. Students may find beneficial a review of the process of reflecting points across symmetry lines, which stresses that a point and its reflected image point are each at the same distance from the symmetry line. As an extension to question 2, students might be asked to connect preimage point *A* and image point *A'* and then describe how the *x*-axis is related to the resulting segment *AA'*. The *x*-axis is a perpendicular bisector of segment *AA'*. A similar question may be posed for other image and preimage points in question 2. The same questions may be posed in question 5.

Sheet 2: In this activity, students are given the same triangle *ABC* as in sheet 1. However, here they translate triangle *ABC* to a series of distorted coordinate axes. The result will be a distorted, or "morphed," image of triangle *ABC*. Grid 1 and grid 2 might be called standard grids because each cell in the grid is congruent and each grid has two sets of parallel grid lines. In grid 1, the coordinate axes are perpendicular and the resulting cells are congruent rectangles. In grid 2, the coordinate axes are not perpendicular and the resulting cells are congruent parallelograms. Grid 1 distorts, or stretches, the triangle along the *x*-axis while shrinking the triangle along the *y*-axis. The result is an image triangle in which neither segment lengths nor angle measures are preserved, except for the angle *A*, the right angle. Grid 2 distorts all segment lengths and angle measures of the original triangle. For both grid 1 and grid 2, the image triangles are all congruent, regardless of the position of triangle *ABC* in the original grid. Grid 3,

grid 4, and grid 5 may be termed *irregular grids* or *nonstandard grids*, since they do not consist of two sets of parallel grid lines. As a result, each grid cell is different. For these three grids, each change in the position of triangle *ABC* in the original grid will produce different image triangles. In each grid, students are actually translating only the vertex points of triangle *ABC* to the distorted grids. Once the vertex points are correctly translated, they can then be joined by segments to produce the image triangles. Since the sides of the image triangles are straight, no need arises to translate other points of triangle *ABC*.

Sheet 3: The activity for this sheet presents students with a cartoon to morph, or translate, onto several distorted grids by using the coordinates to help distort the cartoon face. Since most of the lines in the illustration are curved, students must use the grid lines to help them draw the illustration. Emphasize to students the importance of having the lines of their morphed cartoon face cross the grid lines at the same location as in the original sketch. If the intersection points are accurate, the portion of the sketch between the grid lines is easier to draw. For grid 3, the grid lines are curved, which makes using the intersection points all the more crucial; sketching curved lines on a curved coordinate grid is difficult for many students to do visually without the aid of grid lines. You might ask students to consider how to determine the image coordinates of a point in the cartoon face that is between the grid lines. Since these points are on curved lines, they cannot be located and drawn simply by connecting grid points with segments, as on sheet 2.

After completing sheet 3, students are ready to construct their mirrored cylinders. If you have adhesive reflective Mylar, do not peel off the backing. When the backing is removed, the Mylar will pick up all the surface irregularities of the tube and is a poor reflector. When students tape the reflective Mylar on the tube, suggest that they tape vertically along the overlap seam so that the tape will not affect the reflection.

Sheet 4: In this activity, students explore the reflection properties of the mirrored cylinder. By seeing how the mirrored cylinder distorts basic line relationships, students get a sense of how the mirrored cylinder distorts their artwork. They also gain experience in locating reflections in the mirrored cylinder.

Sheet 5: This sheet contains a normal coordinate grid and a distorted coordinate grid for students

Students explore the reflection properties of the mirrored cylinder

Using Activities from the *Mathematics Teacher* to Support *Principles and Standards*

to use for their anamorphic art. Students first make a sketch, a geometric pattern, or a monogram on the standard coordinate grid. They then transfer their artwork to the distorted grid, just as they did in the activities on sheet 3. In this activity, one set of grid lines consists of arcs of concentric circles that intersect the coordinate axes at uniform intervals. Once students have transferred their original sketches to the distorted grid, have them cut along the heavy horizontal line. The resulting sheet will display their anamorphic artwork with no hint of how they drew the sketch.

Although this activity lies in the area of enrichment, some possible assessments make this activity one that students could include in their portfolios. Grading might be linked to the presentation of the drawing or coloring. Evaluating artistic designs is difficult, since some students have more natural talent than others. For students who are unable to draw something that they find appealing, you might suggest a geometric pattern with vibrant colors, a monogram with stylistic lettering, or a cartoon figure over which the rectangular grid has been drawn. Such alternatives allow even artistically challenged students the opportunity to produce anamorphic artwork. Students might show their distorted sketches to classmates and challenge them to describe the resulting anamorphic art. As an extension of this activity, ask students to build a cone-shaped mirror with their Mylar. When viewed from directly above the vertex, a cone-shaped mirror reflects the full 360 degrees of an anamorphic artwork.

Solutions: Sheet 1:

1. $A(2, 2)$, $B(2, 4)$, $C(6, 2)$

2. $A'(2, -2)$, $B'(2, -4)$, $C'(6, -2)$

3. The y-coordinates of the image, $A'B'C'$, are the opposite of the y-coordinates of the preimage, ABC.

4. $Q'(1, -3)$, $R'(6, -3)$, $S'(5, -5)$. The y-coordinates have the same relationship as in question 3.

5. $A''(-2, 2)$, $B''(-2, 4)$, $C''(-6, 2)$

6. The x-coordinates of the image, $A''B''C''$, are the opposite of the x-coordinates of the preimage (ABC).

7. $Q''(-1, 3)$, $R''(-6, 3)$, $S''(-5, 5)$. The x-coordinates have the same relationship as in question 6.

8. Both reflections result in $A(-2, -2)$, $B(-2, -4)$, $C(-6, -2)$.

9. Reflection across coordinate axes is commutative.

Solutions: Sheet 2:

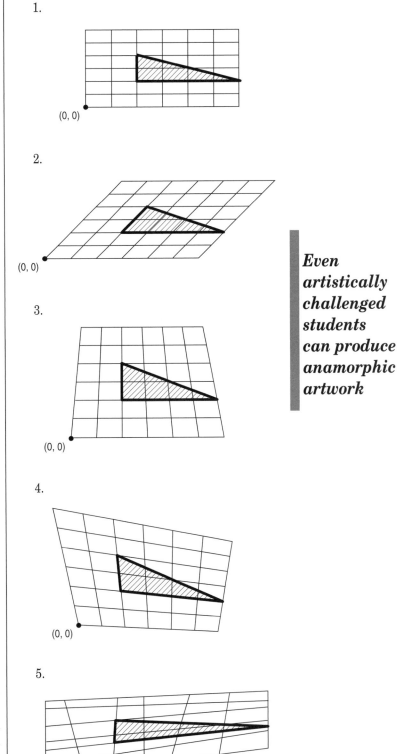

1.

(0, 0)

2.

(0, 0)

3.

(0, 0)

4.

(0, 0)

5.

(0, 0)

Even artistically challenged students can produce anamorphic artwork

Solutions: Sheet 3:

1.

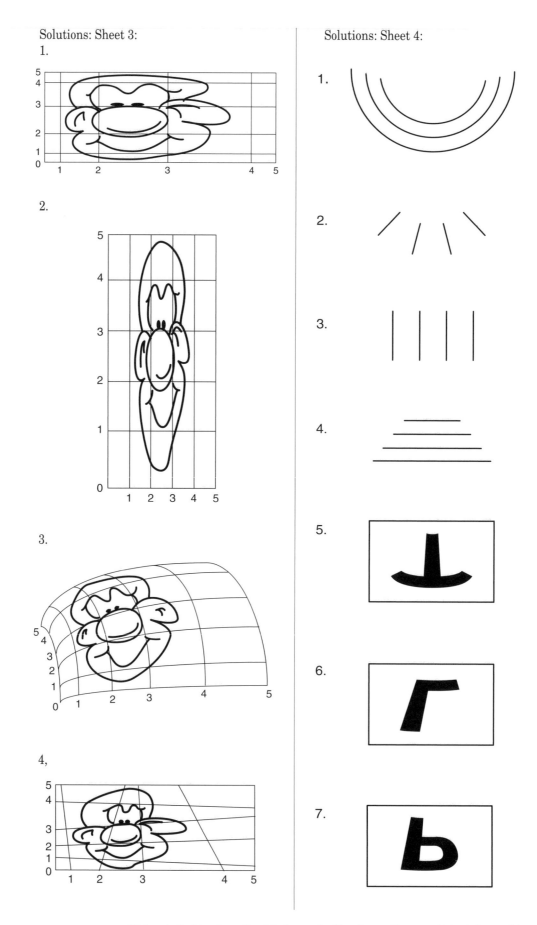

2.

3.

4,

Solutions: Sheet 4:

1.

2.

3.

4.

5.

6.

7.

Using Activities from the *Mathematics Teacher* to Support *Principles and Standards*

BIBLIOGRAPHY

Anno, Mitsumasa, et al. *Anno's Magical ABC: An Anamorphic Alphabet*. New York: Philomel Books, 1981.

Baltruˆsaitis, Jurgis. *Anamorphic Art*. New York: Harry N. Abrams, 1977.

Billings, Karen, et al. *Art 'N' Math*. Eugene, Ore.: Action Math Associates, 1975.

Birmingham, Duncan. *'M' Is for Mirror*. Norfolk, England: Tarquin Publications, 1988.

Bolton, Linda. *Hidden Pictures*. New York: Dial Books, 1993.

Kadesch, Robert. *Math Menagerie*. New York: Harper & Row, 1970.

Leeman, Fred. *Hidden Images: Games of Perception, Anamorphic Art, Illusion*. New York: Harry N. Abrams, 1976.

McLoughlin Bros. *The Magic Mirror: An Optical Toy*. New York: Dover Publications, 1979.

Moscovich, Ivan. *The Magical Cylinder*. Norfolk, England: Tarquin Publications, 1988.

Sandburg, Carl. *Arithmetic*. Singapore: Harcourt Brace Jovanovich, 1993.

Walter, Marion. *The Mirror Puzzle Book*. Norfolk, England: Tarquin Publications, 1985.

1. What are the coordinates of the vertices of triangle *ABC*? _____

2. Reflect triangle *ABC* over the *x*-axis to form image triangle *A'B'C'*. What are the coordinates of the vertices of image triangle *A'B'C'*? _____

3. What relationship do you see between the coordinates of the vertices of *ABC* and the coordinates of the vertices of triangle *A'B'C'*? _____

_____

_____

_____

_____

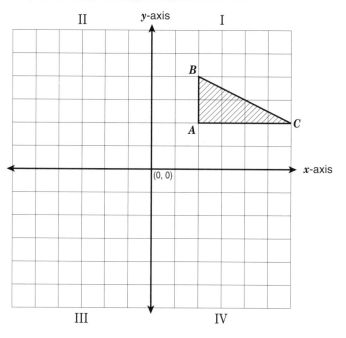

4. Test your relationship from question 3. Use triangle *QRS*, with coordinates *Q*(1, 3), *R*(6, 3), and *S*(5, 5), and reflect triangle *QRS* over the *x*-axis. _____

_____

5. Reflect triangle *ABC* over the *y*-axis to form image triangle *A"B"C"*. What are the coordinates of the vertices of the image triangle *A"B"C"*? _____

_____

6. What relationship do you see between the coordinates of the vertices of triangle *ABC* and the coordinates of the vertices of the image triangle *A"B"C"*? _____

_____

7. Test your relationship from question 6. Use triangle *QRS*, with coordinates *Q*(1, 3), *R*(6, 3), and *S*(5, 5), and reflect the triangle over the *y*-axis. _____

_____

8. Reflect image triangle *A'B'C'* over the *y*-axis, and reflect image triangle *A"B"C"* over the *x*-axis. What do you notice? _____

_____

9. Make a conjecture based on your findings in question 8. Check your conjecture with the image triangle of triangle *QRS*. _____

_____

Graph triangle *ABC* onto each of the following
distorted grids. On the lines beneath each grid,
describe the distorted triangle. Before you begin,
predict which grid will produce the most distorted
triangle. If triangle ABC were translated to a different position on the original grid, it would still
have the same shape. Would the resulting image
triangle in each of these grids still have the same
shape as the first image triangle you graphed?

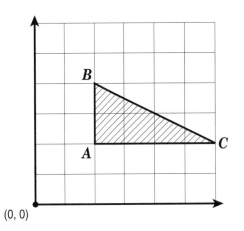

(0, 0)

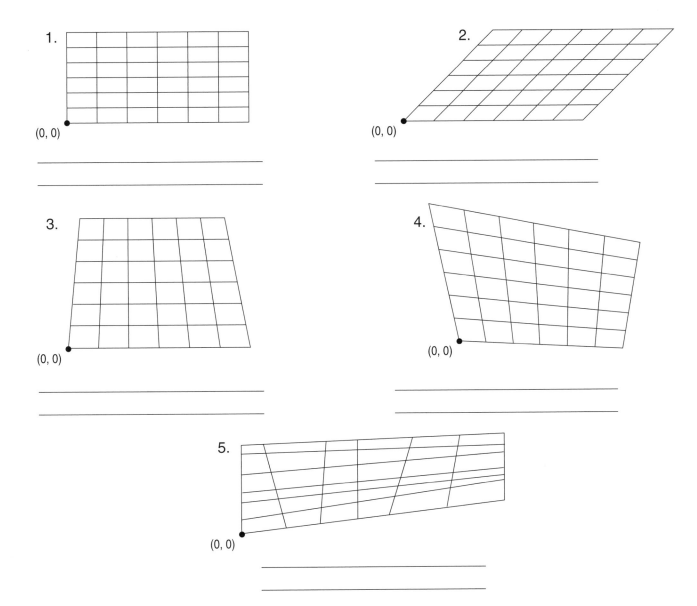

1.

(0, 0)

_____

_____

2.

(0, 0)

_____

_____

3.

(0, 0)

_____

_____

4.

(0, 0)

_____

_____

5.

(0, 0)

_____

_____

You are challenged with graphing a figure that is made of segments and curves. Reproduce the cartoon face on each distorted grid. Be sure that your cartoon lines cross the grid lines at the same place as in the original grid.

Before you begin, predict which graph will produce the most distorted face.

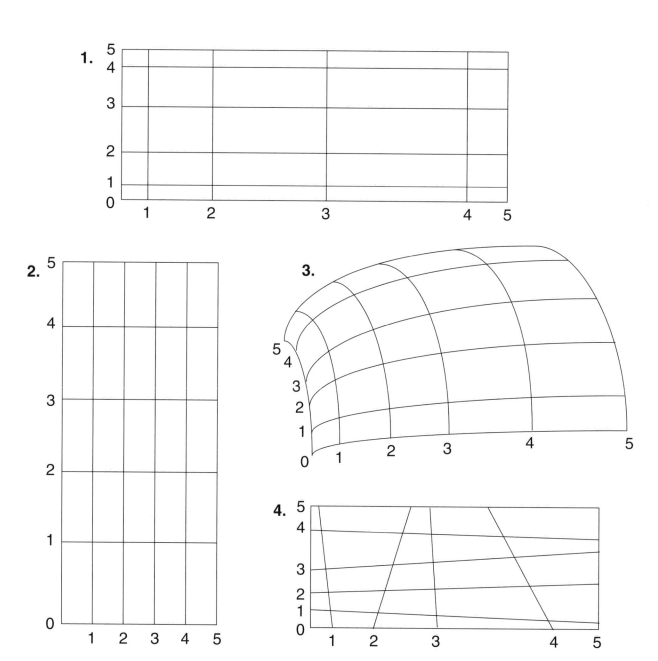

For each of the following, predict what the reflections of the sets of segments or arcs will look like. Draw your prediction.

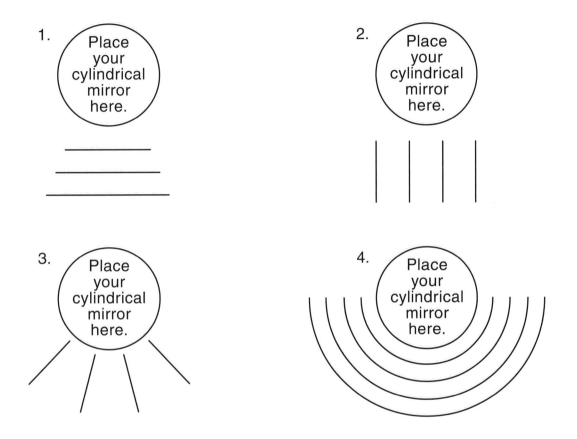

Use what you learned about the reflection of lines to sketch the letters in the boxes so that their reflection in the cylindrical mirror appears with no distortion. Use a pencil so that you can make changes in your sketches.

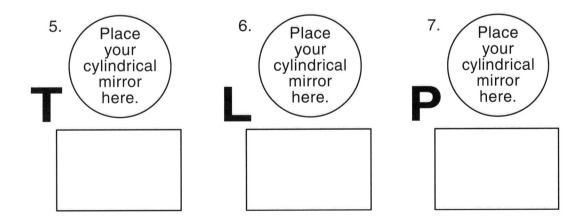

Draw an original sketch in the square grid. Translate your sketch onto the circular grid to form a distorted image. Reflect your distorted image onto a mirrored cylinder to produce a normal image. You may cut along the bold horizontal line to remove the rectangular grid before you display your anamorphic art.

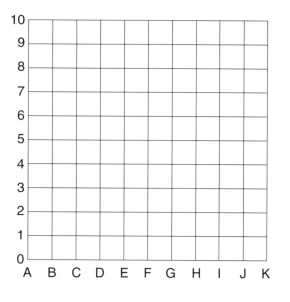

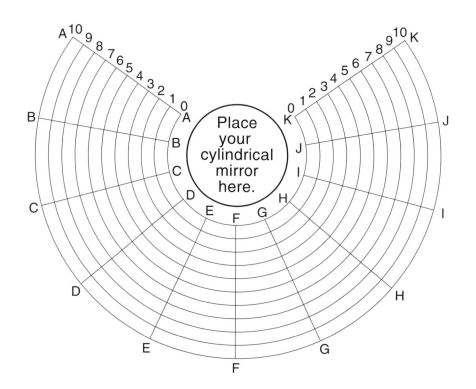

From the *Mathematics Teacher*, January 1998

Probability, Matrices, and Bugs in Trees

Paula Grafton Young   May 1998

Edited by Claudia Carter, *Mississippi School for Mathematics and Science, Columbus, Mississippi*

TEACHER'S GUIDE

The problem

Suppose that three trees are located around a small lake or pond. One of the trees becomes infested with an insect population that destroys the leaves or the fruit of the tree. If the insects randomly move from tree to tree, can we model the path of one insect? Subject to various initial conditions, can we also determine how the insect population will spread among the trees in the long run? Using basic probability, simple random walks, matrices, and Markov chains leads to the answer of "yes" to both questions.

Overview and prerequisites

Inspiration for this exercise came from a section in the "Biogeography" chapter of *Mathematics in Medicine and the Life Sciences* (Hoppensteadt and Peskin 1992). Although this exercise was originally designed for use in a college mathematical-modeling course intended for biology, chemistry, and mathematics majors, the author adapted it for high school students attending a summer program.

Part 1 of the activity incorporates probability and technology into a simple model. Because part 1 requires so little prior knowledge and minimal preliminary discussion (see example 1), it is appropriate for a wide variety of audiences with varied mathematical backgrounds in many different settings. An audience with a strong mathematical background can delve deeper into what is happening with the model or can make up new models.

Part 2 of the activity requires a knowledge of matrix multiplication and elementary probability. The teacher might want to begin discussing Markov chains before presenting this part (see example 2).

The discussion of examples 1 and 2 is presented here as background material from which teachers can draw while introducing the activity sheets to

Paula Young teaches at Salem College, Winston-Salem, North Carolina. She is interested in applications of mathematics to biology and in encouraging students from other majors to pursue mathematics as a major or minor.

students. Allow students to develop these ideas on their own as much as possible.

Materials needed

For part 1, a calculator with a random-number generator is necessary. For part 2, a calculator with matrix-multiplication capabilities is required. For those using an older calculator that cannot handle more than one matrix operation at a time, such as the Texas Instruments TI-81, see Young (1993) for a short program for raising matrices to positive-integer exponents.

An introductory model—the path of one insect

Suppose that three trees are located next to a pond (see fig. 1). One of these trees becomes infested with an insect species that devours the fruit or the leaves of the tree. The bugs randomly move back and forth along the trees until nothing is left for them to eat. We can simulate the path of one insect using *random walks*.

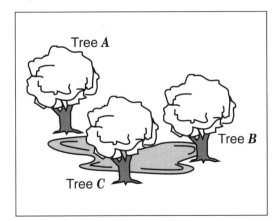

Fig. 1. Three trees located next to a pond

To perform the simulations, we must make some preliminary assumptions:

- Observations are made at regular intervals.

- Between observations, a bug never moves more than one tree away from its previous position.

- A bug will move *clockwise* (cw) between observations with probability p ($0 \leq p \leq 0.5$).

• A bug will move *counterclockwise* (ccw) between observations with the same probability, p.

• A bug will remain on the same tree with probability $1 - 2p$.

• The bug population remains constant; that is, no new bugs move in, the bugs do not reproduce, and no bugs die.

These assumptions simplify the mathematics involved in the model.

We can set up a number line to help us keep track of the probabilities given:

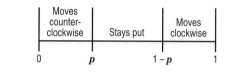

Notice that the lengths of the line segments formed correspond to the probabilities associated with the problem: (1) The distance between 0 and p is the probability that a bug moves to an adjacent tree and corresponds to a bug moving counterclockwise; (2) the distance between p and $1 - p$, which is $1 - 2p$, is the probability that an insect does not move; and (3) the distance between $1 - p$ and 1, which is p, is the probability that an insect moves to the adjacent tree in the clockwise direction. When the random-number generator is used, a number from 0 to 1 will be obtained; that number's position on the number line will tell us whether the insect moves and in which direction.

Example 1. Suppose that the probability that any one of the insects moves to an adjacent tree between observations is $p = 0.3$. Then the probability that an insect does not move will be $1 - 2p$, or 0.4. The chances are therefore pretty good that the insects will move around. Let us simulate one insect's moves through ten observations, if it is initially on tree A.

Procedure. First, put the probabilities on a number line:

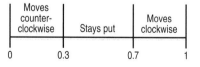

Next, using the random-number generator on a calculator, we can find ten random numbers between 0 and 1. Rounded to two decimal places, the values might be {0.01, 0.42, 0.31, 0.97, 0.03,

0.84, 0.62, 0.21, 0.98, 0.72}. The first number, 0.01, is between 0 and 0.3, so the bug moves counterclockwise from tree A; in other words, it moves to tree C (see fig. 1). The second number, 0.42, is between 0.3 and 07, so the bug stays on tree C. By continuing in this manner, the list of ten values translates into the following path for the insect:

A → C → C → C → A → C → A → A → C → A → B

Introductory model continued—the spread of the population

Here, we use *Markov chains*, that is, matrix multiplication with probability matrices, to determine how the bugs will eventually be distributed through the trees. To estimate the long-term distribution of insects, we must have an *initial state* matrix, which we will call S_0, and a *transition matrix*, which we will call T. The initial state matrix will be a 1×3 matrix, with each entry representing the proportion of the insect population that infests a particular tree. For instance, if 50 percent of the insects are initially on tree A, 20 percent are on tree B, and 30 percent are on tree C, then the initial state matrix would be as follows:

$$\begin{array}{ccc} A & B & C \end{array}$$
$$S_0 = [0.50 \quad 0.20 \quad 0.30]$$

The transition matrix will be a 3×3 matrix, with each row and each column corresponding to one of our trees. Each entry in the matrix represents the probability that an insect on the tree corresponding to the row will move to the tree corresponding to the column between observations.

Example 2. Refer to example 1. The probability that an insect moves to an adjacent tree is 0.30, and the probability that an insect does not move is 0.40. Find the transition matrix, and use the initial state matrix to determine the distribution of the insects at the next observation.

Procedure. Our transition matrix will be as follows:

Next Tree

$$T = \begin{bmatrix} A & B & C \\ 0.4 & 0.3 & 0.3 \\ 0.3 & 0.4 & 0.3 \\ 0.3 & 0.3 & 0.4 \end{bmatrix} \begin{array}{l} A \\ B \text{ Current Tree} \\ C \end{array}$$

Let us learn how this transition matrix, together with the initial state matrix, models the random path of the insect population. If the initial population of insects is distributed as previously indicated, what percent of the population will be on tree A at the next observation? According to our model, 40 percent of those that were initially on tree A will stay there. Therefore, 50% × 40% = 20% of the total insect population will not move from tree A. However, some insects will move from tree B to tree A. Since 20 percent of our initial population was on tree B and 30 percent of these will move counterclockwise to tree A, we have a 20% × 30% = 6% of the total insect population moving from tree B to tree A. Finally, some insects will move clockwise from tree C to tree A. Since 30 percent of the initial population was on tree C and 30 percent of these will move to tree A, we have 30% × 30% = 9% of the total population moving from tree C to tree A. Hence, 20% + 6% + 9% = 35% of the total population will be on tree A at the next observation.

Next, let us look at the product $S_0 \times T$ and compare it with the situation previously given:

$$S_0 \times T = [0.50 \quad 0.20 \quad 0.30] \times \begin{bmatrix} 0.4 & 0.3 & 0.3 \\ 0.3 & 0.4 & 0.3 \\ 0.3 & 0.3 & 0.4 \end{bmatrix}$$

$$= [\mathbf{0.5 \bullet 0.4 + 0.2 \bullet 0.3 + 0.3 \bullet 0.3}$$
$$0.5 \bullet 0.3 + 0.2 \bullet 0.4 + 0.3 \bullet 0.3$$
$$0.5 \bullet 0.3 + 0.2 \bullet 0.3 + 0.3 \bullet 0.4]$$
$$= [\mathbf{0.35} \quad 0.32 \quad 0.33]$$

Observe the bold entry in the product matrix; it corresponds exactly with the arithmetic performed to find the proportion of the population that would be on tree A at the next observation. The second entry, then, corresponds to the proportion of the total population on tree B at the next observation; and the third corresponds to the proportion on tree C at that time. Hence, the product $S_0 \times T$, which we denote by S_1, gives the distribution of the insects at the second observation.

Similarly, to find the distribution of insects in the subsequent observation, S_2, we would calculate as follows:

$$S_2 = S_1 \times T = (S_0 \times T) \times T = S_0 \times T_2$$

The last equality holds by the associativity of matrix multiplication. To find S_3, we would calculate

$$S_3 = S_2 \times T = (S_1 \times T) \times T = (S_0 \times T_2) \times T = S_0 \times T_3.$$

In general, the distribution of insects during the $(n + 1)$st observation is given by

$$S_n = S_0 \times T^n.$$

In our example, if we continue calculating S_n, we see that the state matrix gets closer and closer to

$$\begin{bmatrix} \frac{1}{3} & \frac{1}{3} & \frac{1}{3} \end{bmatrix}$$

so the insects are becoming evenly spread among the trees. If the state matrix were ever actually to reach

$$\begin{bmatrix} \frac{1}{3} & \frac{1}{3} & \frac{1}{3} \end{bmatrix}$$

it would stay that way forever. A state matrix S for which $S \times T = S$ is called a steady state matrix.

Notes about activity sheets

The model presented in parts 1 and 2 of the activity is based on a circle of five trees surrounding a lake (see fig. 2). Part 1 models the path of a single insect under various initial conditions, as in example 1. Before working through part 1, be certain that all students know where to find the random-number generator on their calculators and that the generators are set to produce the desired effect. Most generators can be seeded to ensure that all students obtain the same "random" numbers in the same order. For example, ask students to store "0" as "rand." Then the same sequence of numbers will appear on everyone's screens when executing the generator, provided, of course, that all students are using the same calculator model. If you prefer that all students not obtain the same sequence of values, which is my personal preference, ask students to execute the random-number generator over and over for a minute or two, giving everyone's generator a different starting point. On rare occasions, the random-number generator may produce a number that is exactly equal to p or to $1 - p$. To avoid confusion, one can consider a number in the interval $[0, p]$ to indicate counterclockwise movement, a number in the interval $[1 - p, 1]$ to indicate clockwise movement, and a number in the interval $[p, 1 - p]$ to indicate that no movement occurs.

Observe the bold entry in the product matrix

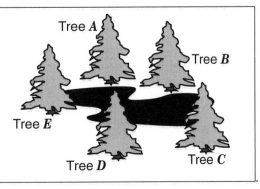

Fig. 2 Circle of five trees surrounding a lake

Part 2 models the spread of the population, as in example 2. To assist with this part, the teacher can use an elementary treatment of Markov chains, which can be found in most finite mathematics textbooks, such as Long and Graening (1997). For a more advanced explanation of Markov chains, see Noble and Daniel (1977) or Scheaffer (1995). For students with strong mathematical backgrounds, additional discussions about the long-term behavior of the model can incorporate the concept of *steady states* and how to find them, giving students a chance to solve equations involving matrices.

> *Students should be encouraged to criticize the model at every stage*

Students should be encouraged to criticize the model at every stage. They should be allowed to offer alternatives to the assumptions made by the model and to incorporate them into a different version of the problem if they see fit to do so.

Conclusion

By using realistic, easy-to-understand models that incorporate different branches of mathematics and that require the use of technology, we give students the opportunity to see mathematics in action. It becomes a hands-on science rather than a group of repetitive exercises. By encouraging students to develop their own models or to criticize existing ones, we give them the opportunity to think, reason, and apply their own knowledge of mathematics, science, and the "real world."

Solutions

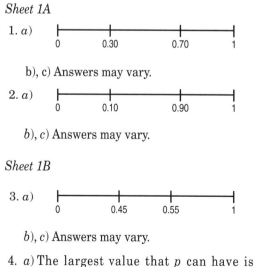

Sheet 1A

1. *a)*

   | | | | |
   | 0 | 0.30 | 0.70 | 1 |

   b), c) Answers may vary.

2. *a)*

   | | | | |
   | 0 | 0.10 | 0.90 | 1 |

   b), c) Answers may vary.

Sheet 1B

3. *a)*

   | | | | |
   | 0 | 0.45 | 0.55 | 1 |

   b), c) Answers may vary.

4. *a)* The largest value that p can have is 0.50, because we are assuming that an insect moves left or right with the same probability.

b) The corresponding value of $1 - 2p$ is then 0, meaning that an insect never stays put; it always moves left or right.

5. *a)* The value of p for tree C is 0.50.

   b) When running the simulation, the same rules apply to trees A, B, D, and E, as before. Only when an insect lands on tree C is the new assumption used. Answers will vary as before.

6. *a)* The value of p is now 0 for tree C.

   b) See 5 (*b*).

Sheet 2A

7. Students should notice that the distribution of insects approaches a limit; the bugs will eventually be evenly distributed among the trees.

8. Students should again notice that the distribution of insects is approaching a limit; the bugs are eventually evenly distributed among the trees.

Sheet 2B

9. Again, the bugs are evenly distributed among the trees. Students should also notice that the proportion of insects on each tree oscillates before reaching 20 percent.

10. This time, tree C has very few bugs. Unless the insects remember that tree C is barren, a few insects will always move from trees B and D to tree C. However, no insect ever stays there.

Sheet 2C

11. Very slowly, all insects are moving into tree C and staying there.

References

Hoppensteadt, Frank C., and Charles S. Peskin. *Mathematics in Medicine and the Life Sciences.* New York: Springer-Verlag, 1992.

Long, Paul A., and Jay Graening. *Finite Mathematics: An Applied Approach.* 2d ed. Reading, Mass.: Addison Wesley Longman Publishing Co., 1997.

Noble, Ben, and James W. Daniel. A*pplied Linear Algebra.* 2d. ed. Englewood Cliffs, N. J.: Prentice-Hall, 1977.

Scheaffer, Richard L. Introduction to *Probability and Its Applications.* 2d. ed. Pacific Grove, Calif.: Brooks-Cole Publishing, Duxbury Press, 1995.

Young, Paula G. *Graphing Calculator Lessons for Finite Mathematics.* New York: HarperCollins College Publishers, 1993.

PART 1

Use the following diagram for reference throughout the activity.

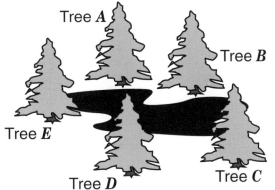

Tree *A*
Tree *B*
Tree *E*
Tree *C*
Tree *D*

1. Suppose that the probability that any one of the insects moves left or right to an adjacent tree between observations is $p = 0.3$. Then the probability that an insect does not move will be $1 - 2p = 0.4$. In this exercise, we simulate how one insect that is initially on tree A moves through ten observation periods.

   a) Label the number line with the appropriate probabilities.

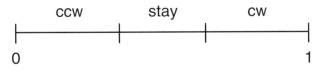

ccw   stay   cw

0   1

   b) Generate ten random numbers on your calculator, and list them in order. (Round to two decimal places.)

   c) Translate these random numbers into the path of the bug:

   A →____→____→____→____→____→____→____→____→____→____.

2. Suppose that the probability that a particular insect moves to an adjacent tree between observations is $p = 0.10$ and that the insect is initially on tree D.

   a) Label the number line with the appropriate probabilities.

ccw   stay   cw

0   1

   b) Generate ten random numbers, and list them in order.

   c) Translate your list of random numbers into a simulated insect path through the circle of trees.

   D →____→____→____→____→____→____→____→____→____→____.

From the *Mathematics Teacher*, May 1998

3. Repeat question 2 for $p = 0.45$.

   a)

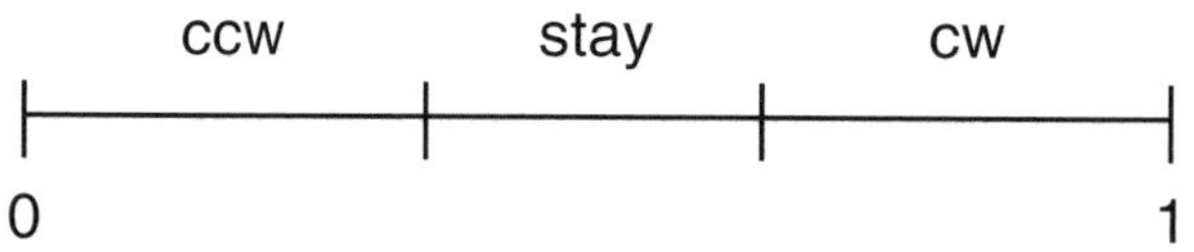

ccw   stay   cw

   ├————————————┼————————┼——————————┤

   0   1

   b) Random Numbers

   ____   ____   ____   ____   ____   ____   ____   ____   ____   ____

   c) Path of the Bug

   D →____→____→____→____→____→____→____→____→____→____.

4. a) What is the largest value that p can have in a simulation of this type? _____

   Why? _____

   b) What is the corresponding value of $1 - 2p$? _____ What does this value mean
      about the path of an insect in this model?

5. Suppose that tree C has a disease that prevents it from bearing any fruit. Thus, an
   insect that finds itself on tree C does not want to stay there. The probability that an
   insect stays on tree C through the next observation is therefore 0.

   a) What is the value of p for tree C? _____

   b) Run a simulation using the probabilities from question 1 for trees A, B, D, and E and
      this new assumption about tree C. Assume that the bug starts out on tree C.

   Random Numbers

   ____   ____   ____   ____   ____   ____   ____   ____   ____   ____

   Path of the Bug

   C →____→____→____→____→____→____→____→____→____→____.

6. Next, suppose that tree C is so covered with fruit that a bug that lands there will never
   want to leave! Thus, the probability that an insect stays on tree C is 1.

   a) What is the value of p for tree C? _____

   b) Run a simulation using the probabilities from question 1 for trees A, B, D, and E and
      this new assumption about tree C. Assume that the bug starts out on tree B.

   Random Numbers

   ____   ____   ____   ____   ____   ____   ____   ____   ____   ____

   Path of the Bug

   C →____→____→____→____→____→____→____→____→____→____.

PART 2 SHEET 2A

7. Suppose that 50 percent of the insects are initially on tree A, 20 percent are on tree B, and 30 percent are on tree E; then the initial state matrix would be this:

$$A \quad B \quad C \quad D \quad E$$
$$S_0 = [\ 0.50 \quad 0.20 \quad 0 \quad 0 \quad 0.30\ ]$$

Suppose that the probability that an insect moves to an adjacent tree is 0.30 and that the probability that an insect does not move is 0.40. Then the transition matrix will be as follows:

Next Tree

	A	B	C	D	E	
	0.4	0.3	0	0	0.3	A
	0.3	0.4	0.3	0	0	B
T =	0	0.3	0.4	0.3	0	C Current Tree
	0	0	0.3	0.4	0.3	D
	0.3	0	0	0.3	0.4	E

Find S_1, S_2, S_5, S_{10}, S_{25}, S_{50}, S_{51}, and S_{52}. What do you notice? _____

_____

_____

_____

_____

8. Consider the transition matrix from question 7.

a) If all insects are initially on tree D, set up the initial state matrix.

$$S_0 = [\ \underline{\quad} \quad \underline{\quad} \quad \underline{\quad} \quad \underline{\quad} \quad \underline{\quad}\ ]$$

b) Find S_1, S_2, S_3, S_4, S_5, S_{10}, S_{20}, S_{25}, S_{50}, S_{51}, and S_{52}. What do you notice?

_____

_____

_____

_____

From the *Mathematics Teacher*, May 1998

9. a) Set up the transition matrix for question 3 in part 1, sheet 1B.

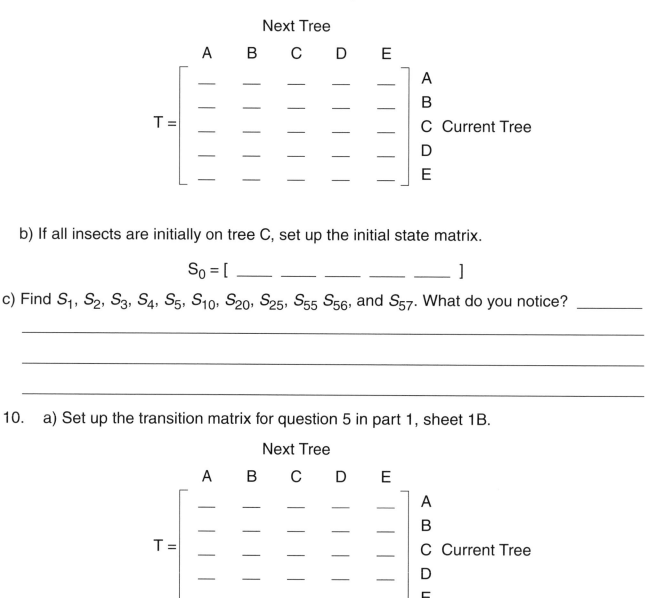

Next Tree

$$T = \begin{bmatrix} \_ & \_ & \_ & \_ & \_ \\ \_ & \_ & \_ & \_ & \_ \\ \_ & \_ & \_ & \_ & \_ \\ \_ & \_ & \_ & \_ & \_ \\ \_ & \_ & \_ & \_ & \_ \end{bmatrix} \begin{matrix} A \\ B \\ C \\ D \\ E \end{matrix}$$

Current Tree

b) If all insects are initially on tree C, set up the initial state matrix.

$$S_0 = [\ \_\_\_ \ \ \_\_\_ \ \ \_\_\_ \ \ \_\_\_ \ \ \_\_\_ \ ]$$

c) Find S_1, S_2, S_3, S_4, S_5, S_{10}, S_{20}, S_{25}, S_{55} S_{56}, and S_{57}. What do you notice? _____

_____

_____

_____

10. a) Set up the transition matrix for question 5 in part 1, sheet 1B.

Next Tree

$$T = \begin{bmatrix} \_ & \_ & \_ & \_ & \_ \\ \_ & \_ & \_ & \_ & \_ \\ \_ & \_ & \_ & \_ & \_ \\ \_ & \_ & \_ & \_ & \_ \\ \_ & \_ & \_ & \_ & \_ \end{bmatrix} \begin{matrix} A \\ B \\ C \\ D \\ E \end{matrix}$$

Current Tree

b) If 25 percent of the insects are initially on tree A, 10 percent are on tree B, and 65 percent are on tree D, set up the initial state matrix.

$$S_0 = [\ \_\_\_ \ \ \_\_\_ \ \ \_\_\_ \ \ \_\_\_ \ \ \_\_\_ \ ]$$

c) Find S_{10}, S_{20}, S_{25}, S_{50}, S_{51}, and S_{52}. What do you notice? Does this result make sense? _____

_____

_____

_____

11.a) Set up the transition matrix for exercise 6 in part 1, sheet 1B.

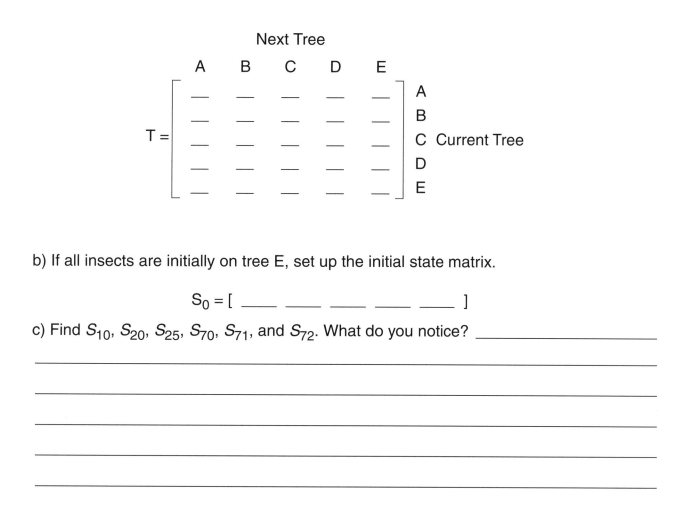

b) If all insects are initially on tree E, set up the initial state matrix.

$$S_0 = [\underline{\quad} \quad \underline{\quad} \quad \underline{\quad} \quad \underline{\quad} \quad \underline{\quad} \ ]$$

c) Find S_{10}, S_{20}, S_{25}, S_{70}, S_{71}, and S_{72}. What do you notice? _____

_____

_____

_____

_____

_____

Extra Problems

1. This simulation does not consider much information about the insects or about the trees. For example, in the original model, we have not considered what the insects do if all the fruit and leaves are gone. Make a list of all the information that the model does not consider.

2. Suppose that the trees are in a row rather than in a circle. Develop a list of assumptions, similar to those in part 1, sheet 1, that could be used to model the path of the insects.

3. Develop your own model using the original assumptions and some of the items that you listed in the first extra problem. Run some simulations, and share them with the class. Describe what happens in the long run, and offer explanations.

Forest Fires, Oil Spills, and Fractal Geometry: An Investigation in Two Parts

L. Charles Biehl

November 1998

Edited by Claudia Carter, *Mississippi School for Mathematics and Science, Columbus, Mississippi*

Part 1: Cellular Automata and Modeling Natural Phenomena

Teacher's Guide

Complex models can be generated with a simple set of rules

The burn pattern of a forest fire, the percolation pattern of an oil spill seeping into the earth, and the area over which a disease spreads—all are complex in shape and structure. An analysis of these patterns can provide researchers with information that they need to better understand the nature and causes of these phenomena. In this set of activities, students explore how complex models can be generated with a small and simple set of rules. In the second article of this two-part series, students will analyze these models while they investigate the relationship between mathematical models and the phenomena that they represent. The second article, "Using Fractal Complexity to Analyze Mathematical Models: More on Forest Fires and Oil Spills," will be published in the February 1999 issue of the *Mathematics Teacher.*

Cellular automata (CAs) are the graphical equivalent of a fleet of tiny, simple computer programs, each capable of performing the same single simple task. These programs are executed in the form of rows of cells arranged in a grid pattern, and the task that they perform is simply to shade or not to shade the cells in the next row, on the basis of a single rule. In the simplest situation, called a *one-dimensional CA,* the cells arranged in a row apply their rule to determine whether their neighbors in the next row will be shaded or blank. A corresponding numerical example is Pascal's triangle, in which each pair of numbers in a row determines the value of its neighbor in the next row. In Pascal's triangle, the rule is that the sum of the two adjacent numbers in a row becomes the value of the cell below and between them in the next row. (See fig. 1.) Rather than be numbered, CA cells are simply shaded or not shaded, depending on the cells in the previous row, according to a rule that can be expressed as a verbal or visual algorithm.

The cells of two-dimensional CAs are arranged in a plane and are shaded by rules that use neighbors in two dimensions. CAs can also exist in three or

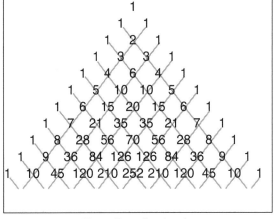

Fig. 1. Pascal's triangle

more dimensions. The best-known example of a two-dimensional CA is the "game of life," created by John H. Conway, a mathematician at Princeton University (Berlekamp, Conway, and Guy 1982). In the "game of life," the cells are arranged in a rectangular grid. A set of shaded cells, the live ones, called the *seed,* is placed on the grid. Following a predetermined set of rules, cells will "live" or "die" depending on the number and configuration of neighbors of each cell. Some cells get starved out, some die of loneliness, and some reproduce wildly, resulting in interesting and complex patterns. Much work has been done with the "game of life"; several Internet URLs given at the end of this activity allow students to explore the game and even download software to demonstrate it.

Much research has been done using CAs, primarily because they "exhibit behaviors and illustrate concepts that are unmistakably physical" (Vichniac 1984, 96). Studies of CAs have furnished useful information for digital filters and the growth

L. Charles Biehl teaches at The Charter School of Wilmington, Wilmington, Delaware. His interests include bringing discrete mathematics, especially twentieth-century mathematics, into the classroom.

Editorial Comment: On Sheet 4, students had difficulty distinguishing between newly ignited trees and those burning already. One strategy they developed was to alternate pen colors for each round. A burning tree was a dot; a burned out tree was shaded. A teacher may wish to model a forest fire with the class to elicit students' suggestions before groups begin work. See page 139 for Part 2 of this activity.

of crystals. "One of the motivations for studying cellular automata in general is to gain insight into the nonlinear phenomena that occur in the solution of differential equations, and in the physical systems they are used to model" (Park, Steiglitz, and Thurston 1986, 431). In plain language, the use of CA models can demonstrate many patterns seen in nature or in man-made environments. From such simple games as John Conway's "game of life" to the patterns seen in the chapter headings of Michael Crichton's book *The Lost World* (1995), CA models use simple sets of rules to produce images that are both complex and beautiful. These images can demonstrate such widely diverse ideas as reproduction and growth of populations, from amoebas and cancer cells to the growth of urban centers, as well as the spreading patterns of lichens and wildflowers. Many experiments use CAs to study the behaviors of particles as they interact, as shown in the one-dimensional CA in figure 2.

CA models can demonstrate many patterns in a man-made environment

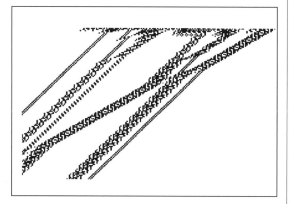

Fig. 2. A cellular-automata model showing interaction among cells. From a monograph by James K. Park, Kenneth Steiglitz, and William P. Thurston; used with permission

Students begin these activities by experimenting with a one-dimensional CA model. This exercise is followed by an activity involving a variation of a one-dimensional CA model called *oriented percolation* to generate models of the profile of an oil spill. The final activity uses two-dimensional CAs to model burn patterns of forest fires.

Grade levels: 9–12

Objectives: By constructing these models, students will use algorithms and probability to understand how a simple iterative procedure can produce complex results. Further, students will investigate the use of mathematical modeling in generating visual representations of physical phenomena.

Materials: Black and red pencils; copies of student handouts, including the introduction, one copy of sheets 1, 3, and 4, and two copies of sheet 2; and a means of generating random numbers, such as a calculator with a random-number generator, percent dice, and so forth. If none of these sources is available, consult sheet 3A for a list of random digits.

Prerequisites: Experience in generating random numbers as single digits, pairs, and triple digits

Directions:

Sheet 1 introduces one-dimensional CAs. Sheet 2 provides an additional exploration based on the same set of rules. Sheet 3 uses CA to generate a model of an oil spill. Sheet 4 models forest-fire-burn patterns.

Sheet 1

This sheet, which introduces one-dimensional CAs, is designed to be completed by individual students. Students must realize that the hexagonal cells are arranged in horizontal rows. Except for the top-row cells that contain the seed, the cells in each row will be shaded or not depending on their neighbors in the previous row. In this example, a cell's state is determined by its two neighbors located just above, one on the left and one on the right. As shown on sheet 2, other rules can be used to determine a cell's state. The seed for this activity is the single shaded cell in the top row.

The rules for shading are given in the form of a lookup table located on the sheet. You might ask the students to tell you what the lookup table says. They should answer that if two consecutive cells in a row are both shaded or both blank, their neighbor below and between them will remain blank; if a pair of consecutive cells in a row consists of one shaded cell and one blank cell, their neighbor below and between them will be shaded. For example, all the cells in the second row will remain blank except for the two directly below the seed, since in the first row, the shaded cell's left and right neighbors form pairs with it that cause the cells immediately below them to be shaded. Figure 3 shows the first few rows. Students must work row by row. Although definite patterns will begin to emerge, be sure that they use the lookup table rather than rely on the patterns that they perceive. After students have filled in a couple of rows, you might suggest that they compare results with those of someone else and discuss any differences to be sure that they are following the rules correctly. Allow time for students to complete all the rows

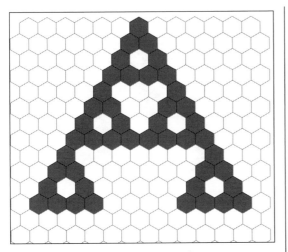

Fig. 3. The first twelve rows of sheet 1

given, and then discuss the results. A definite pattern begins to emerge; students already familiar with introductory fractal geometry may recognize the beginning of the Sierpinski triangle, an often-studied fractal shape. Questions to raise include the following:

- How long can this process continue?
- What would the "finished" CA look like?
- Can it ever be truly "finished"?
- What properties can you see in the pattern?

Students could continue completing rows indefinitely, but several distinct properties can be observed as the figure grows. Try to elicit a list of as many patterns as possible; all conjectures are welcome. Be certain that students understand the process completely before starting sheet 2.

Sheet 2

Each student should have two copies of this sheet. For this activity, students should consider two different aspects. First, what is the result if the same lookup table is used with a different seed? Have students create seeds of three to five cells and then compare results when they finish. They will find that the pattern eventually begins to resemble the one that they saw on sheet 1. The fundamental question is *why*? The answer is that all possible seeds of three to five cells are represented somewhere as part or all of a row on sheet 1. Have students find their seeds somewhere on sheet 1 to confirm this statement, and then ask what would happen with larger seeds. Given sufficient time and materials, this exercise could make an interesting ongoing investigation, especially since the author does not have clear evidence of the

possible results. What happens if students use a different lookup table? Will the same thing happen again (Peitgen, Jürgens, and Saupe 1992)? Answering these questions is the second aspect of the activity. On another copy of the grid, have students design and implement different lookup tables. As previously mentioned, a cell's state can be determined by more than two of its neighbors. Suggest rules that use three or more neighbors. Examples include using four neighbors in the previous row, shading the cell if any three of the four are shaded, if an odd number of them are shaded, if an even number of them are shaded, and so on. Just make sure that students develop lookup tables that use only the single row above.

Assessment. After students have completed the second part of this activity, ask them to write an explanation of everything they see in their own work and in the work of others, especially including patterns that develop, as well as the similarities and differences in the results obtained from different, but possibly related, lookup tables. Students should summarize their understanding of one-dimensional CAs before moving on. Incidentally, these pictures, accompanied by their explanations, also make interesting work to display.

Sheet 3

This activity introduces a little chaos to CAs. In the preceding activities, the cells always react solely to information given in the lookup table. In variations of CAs called *interactive particle systems* and *oriented percolation* (as here), the rule for developing patterns involves an element of randomness that is based on assigning a probability that the cells will react to the lookup table.

Sheet 3 represents the profile of the spill of oil or other liquid percolating into the earth. The cells are arranged in a bricklike pattern to represent particles of soil or other material in the ground; the cells represent the spaces around those particles into which the oil can seep; the soil particles are the intersections where a cell meets its two neighbors in the next row. Cells are considered neighbors if they are in consecutive rows and share part of a face. For a predetermined probability that models the density of the soil, the oil can flow around the soil particles to the left or to the right, but it must always flow downward because of gravity. The bottom edges of each cell act as valves through which the oil seeps, with each valve having a certain probability of being open when the oil reaches that point from above. If the soil bed is porous enough, that is, if the valves have a higher

This activity introduces a little chaos to CAs

probability of being open, the oil will be able to percolate, or flow, through the network of cells, in the form of pools or even "fingers," or flowing strands of liquid, through the soil bed. As the oil percolates through the soil bed, parts of the percolation will end, and depending on the percolation probability, the oil may eventually stop percolating altogether.

The seed for this model is a set of shaded cells in the top row, representing an oil spill as seen from the side. The equivalent of a lookup table in words is as follows: if a cell is shaded, it has a certain predetermined probability of causing each of its two neighbors in the next row to be shaded. Students should work in pairs. Assign various pairs of students the probabilities of 0.5, 0.55, 0.6, 0.65, 0.7, and 0.75. See the "analysis of the results" for the reason for selecting these probabilities. Ask students to develop their own methods for using the random-digit table for their assigned probability. For example, to use a probability of 0.6, the digits 1–6 could indicate percolation, whereas 7–0, using "0" as "10," could indicate no percolation. For the probability of 0.55, read the digits in pairs; 01–55 indicate percolation, whereas 56–00, using "00" as "100," indicate no percolation. Students can also use a calculator with a random-number function to generate digits.

Each member of the student pair has a specific responsibility; one student should find and announce the random number for each cell to indicate whether it gets shaded, and the other should do the actual shading. After a few false starts and communication problems, the students should be able to work together efficiently to shade cells. Since each cell has two neighbors in the previous row, it may have two chances to get shaded, but if it gets shaded by the left neighbor above it, testing it again is unnecessary. Allow at least fifteen minutes for this activity, so that most of the models can be completed. Higher probability models may take longer, but the "fingers" of percolation should be clearly indicated (fig. 4). Students can attach a second sheet of cells to the first one to continue models that run off the page.

Analysis of the results. Experimentation has determined that real percolation, where free flow does not cease, begins with probabilities near 0.645 (Kaye 1989). No proof has yet been offered that 0.645 is a critical value for such percolation. The best proof narrows the bounds for this critical value to somewhere between 0.6298 and 0.84. Students' results will vary; for very low or high probabilities, the model's results will be relatively smooth, since the oil will either not flow at all for

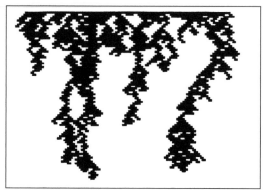

Fig. 4. A fully percolated spill using $p = 0.6$

low probabilities or will flow relatively freely for higher probabilities. For probabilities in the middle of the range, a definite connection exists between the probability of percolation and the characteristic "roughness" of the completed model. The greatest limitation of this model is the fact that the "density" of the soil bed remains constant throughout. More complex models could include pockets of soil that are associated with higher or lower percolation probabilities. In the real world, this assumption of constancy limits the ability to predict percolation damage caused by spills. When various soil and core samples are used to approximate the density of the substrata of the soil bed, appropriate steps are more easily taken in anticipation of reclaiming polluted soil following a spill.

Assessment. After completing their models, student teams should compare their results with those obtained by others, looking for a relationship between the probability used and the resulting image. Questions that can be asked include the following:

- What relationship do you see between the probabilities used and the resulting models?
- How could this model be modified to account for variations in soil density?
- How could having such an image of a real oil spill aid in its cleanup?

Sheet 4

This activity allows students to try their hand at two-dimensional CAs. Students cannot work row by row; the seed is at the center and grows outward in all directions. The cells are arranged in a square grid, with each cell representing a living tree. In a real forest, the trees have neighbors in all directions, and the trees are not uniformly distributed. For the purposes of this model, however, trees have only horizontal and vertical neighbors and are uniformly distributed. Cells that are adjacent diagonally are not considered to be neighbors.

In this model, a cell, which represents a tree, can be in one of three states: unburned, burning, or dead. A burning tree will burn for one "turn"; then it becomes dead. A dead tree stays dead; it cannot reignite. Since this CA model is working in two dimensions, a turn consists of examining burning trees in a systematic manner to see whether they ignite their neighbors then changing them into dead ones. A systematic way of examining the cells is to begin at the top and proceed clockwise around the perimeter of the fire, testing and shading cells along the way.

Students should be grouped in teams of three for this activity. Each team member has one of the following responsibilities: one student is in charge of the random numbers; one, in charge of setting trees on fire; and one, in charge of turning burning trees into dead ones. Assign various groups of students the probabilities of 0.4, 0.45, 0.5, 0.55, and 0.6. Tell them to shade a small cluster of cells close to the center of the grid then use random numbers, as in the oil-spill activity, to run the model as previously described.

Analysis of the results. To the author's knowledge, this model has not been researched as deeply as the oil-spill model. With low probabilities, that is, those below 0.4, the fire tends to burn out rather quickly; with high probabilities, that is, 0.6 or higher, the fire spreads relatively freely, yielding "smoother" burn patterns. In the critical range of 0.4 to 0.6, the complexity of the burn pattern increases then decreases again as the fire becomes more intense. This topic is the focus of the second article in this series. Experimentation with the model and awareness of a relationship between the numbers and the pictures are the goals of this activity (fig. 5).

Assessment. Student groups should compare work to seek patterns, specifically the connection between the probability that they used and the resulting burn perimeter. Questions to ask include the following:

- What would the burn pattern look like as the probability approaches 1.0?

- What properties do the burn patterns exhibit as the probability increases from 0.4?

- How could this model be modified to account for such variables as wind or changes in the density of the trees? To

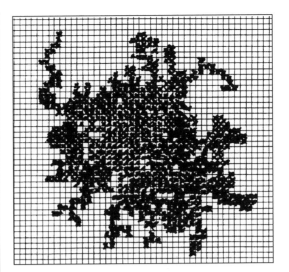

Fig. 5. Forest-fire model using cellular automata

consider wind, increase the windward and decrease the leeward probabilities; designate certain regions to have a different probability to model changes in the density of the trees.

- How would the results differ if a hexagonal grid is used? The author's experience indicates that no significant differences exist; using a hexagonal grid produces fires that are more circular as the probability approaches 1.0.

Conclusions

This very simple process produces some truly complex results. Further, when randomness is introduced, the results become strikingly similar to naturally occurring phenomena. By having students compare their work with that of others, they can discover a correlation between the value of the probability used to generate the model and the physical characteristics of the finished model. So that students can clearly see this relationship, they should generate as many sample models as possible, including many with probabilities between the ones used in these activities. Interested students can create additional samples outside of class; a series of models can be displayed in increasing order of probability.

The overall conclusion from generating oil spills and forest fires is that for some critical range of probabilities, a definite connection exists between the numbers and the complexity of the pictures. The outlines of both models tend to show marked differences in jaggedness and irregularity. A means of quantifying this degree of irregularity is known as the *complexity measure*, sometimes

A very simple process can produce truly complex results

called *fractal complexity*, or even *fractal dimension*, which is the focus of the second article in this series.

Determining mathematical relationships through modeling is a powerful tool in the real world. The ability to take core samples of real spills allows professionals to draw conclusions about the severity of the spill, especially since the outline generated through the modeling process would not be available at an actual spill site. Quantifying the characteristics of real spills by using probabilities and complexities enables professionals to develop guidelines for dealing with them. Similarly generated models of forest-fire burn patterns have been used to provide early intervention information in actual fires.

burn patterns have been used to provide early intervention

Complexity measures of models of crystal growth have been useful in understanding many new materials in the manufacturing field; they are used to measure the roughness of cosmetics, the solubility of soap powders, the porosity of safety-filter masks, and the breakability of cookies. Cellular-automata processes and related iterative processes may soon be used to understand the underlying structure of such complex systems as clouds, riverbeds, and lava flows. The ecology of swamplands has already been analyzed, and of course, so have percolation systems. Cellular automata were not studied in the first half of this century, nor was fractal geometry for the first two-thirds. Both fields are beginning to be used heavily in pure and applied mathematics research in the continued search for applications, which continue to emerge on a regular basis.

Internet Sites Featuring Cellular Automata

- Green, David; Tutorial Notes on Cellular Automata
  (complex.csu.edu.au/complex/tutorials/tutorial1.html)

- Griffeath, David; Cellular Automata Java Applets, University of Wisconsin
  (psoup.math.wisc.edu/java/java.html)

- Holmes, Rich; Links to Cellular Automata Games
  (web.syr.edu/~rsholmes/games/cell_aut)

- Java Resource Center for Cellular Automata and the Game of Life
  (www.javacats.com/US/)

- Rucker, Rudy; CelLab: Cellular Automata Software with Demonstrations and Examples
  (www.mathcs.sjsu.edu/faculty/rucker/cellab.htm)

- Web Interactive Cellular Automata Group
  (www.aridolan.com/wica)

References

Berlekamp, Elwyn R., John H. Conway, and Richard K. Guy. "What Is Life?" In *Games in Particular. Vol. 2, Winning Ways for Your Mathematical Plays.* New York: Academic Press, 1982.

Crichton, Michael. *The Lost World*: A Novel. New York: Alfred A. Knopf, 1995.

Kaye, Brian H. *A Random Walk through Fractal Dimensions.* New York: VCH Publishers, 1989.

Park, James K., Kenneth Steiglitz, and William P. Thurston. "Soliton-like Behavior in Automata." *Physica* 19 (1986): 423–32. Reprinted in *Theory and Applications of Cellular Automata,* edited by Stephen Wolfram, 333–42. Hong Kong: World Scientific Publishing Co., 1986.

Peitgen, Heinz-Otto, Harmut Jürgens, and Dietmar Saupe. *Introduction to Fractals and Chaos.* Vol. 1, *Fractals for the Classroom.* New York: Springer-Verlag, and Reston, Va.: National Council of Teachers of Mathematics, 1992.

Vichniac, Gérard Y. "Simulating Physics with Cellular Automata." In *Cellular Automata: Proceedings of an Interdisciplinary Workshop, Los Alamos, New Mexico 87545, USA, March 7–11, 1983,* edited by Doyne Farmer, Tommasso Toffoli, and Stephen Wolfram, 96–116. New York: North-Holland Physics Publishers, 1984.

CELLULAR AUTOMATA: AN INTRODUCTION FOR STUDENTS

What do the burn patterns of forest fires, the percolation patterns of pollutants seeping into the earth, and the area over which a disease spreads have in common? All are complex images that can be modeled mathematically using simple rules and a dynamic process called *iteration*. In the past few years, the study of chaos theory has led scientists from many fields to conclude that even the most chaotic and random-seeming images contain underlying structure. Major breakthroughs have occurred in such widely varied areas as predicting the onset of heart attacks; understanding the stock market; compressing digital images for transmission; creating extremely complex computer images of trees and flowers; and actually modeling a computer version of artificial life, complete with mutations.

One process for generating and understanding some of these models is called *cellular automata* (pronounced "aw-TAH-ma-ta"), or *CAs* for short, the graphical equivalent of a fleet of tiny, simple computer programs, each capable of performing the same single simple task. These programs are executed in the form of cells, arranged on a grid, that apply a rule to determine whether their neighbors will be shaded or blank. CAs were first used as a theoretical tool in the late 1940s by such mathematicians as John Von Neumann; beginning in the late 1960s, CAs began to find their way into practical applications, such as those in these activities. When arranged in a series of lines or rows, these cells can represent generations of simple life forms. In two dimensions, CAs can also model the evolution of life forms, as in John Conway's "game of life," as well as such wildly diverse phenomena as population growth in amoebas, urban sprawl, or the spreading patterns of cancer cells or patches of wildflowers.

The following set of activities gives you an opportunity to explore the properties of CAs in both one and two dimensions. By using probability to introduce chaos into the process, you can produce and analyze your own models of forest-fire-burn patterns and oil-spill-percolation patterns. These activities will help you understand how to use mathematics to produce meaningful models of natural phenomena, to discuss and analyze these models, and to develop conjectures and reach conclusions. If you wish, you can further pursue this area on your own. Many resources that discuss and investigate CAs and mathematical modeling are available.

Consider the grid of rows of hexagonal cells below, and notice that in each row, each cell has two neighbors in the previous row. The *lookup table* in the upper-right corner shows the rule that the cells in each row follow to decide whether the neighbor below and between them will be shaded or left blank. The shaded cell in the top row is called the *seed*, or starting value, for the cellular automata. Working one row at a time, use the lookup table to shade the appropriate cells in each row.

How can you describe the rule in the lookup table in words? While you shade the cells in each row, look for developing patterns, but follow the lookup table strictly, continuing to work row by row. When you get to the last row, be prepared to describe and share your results.

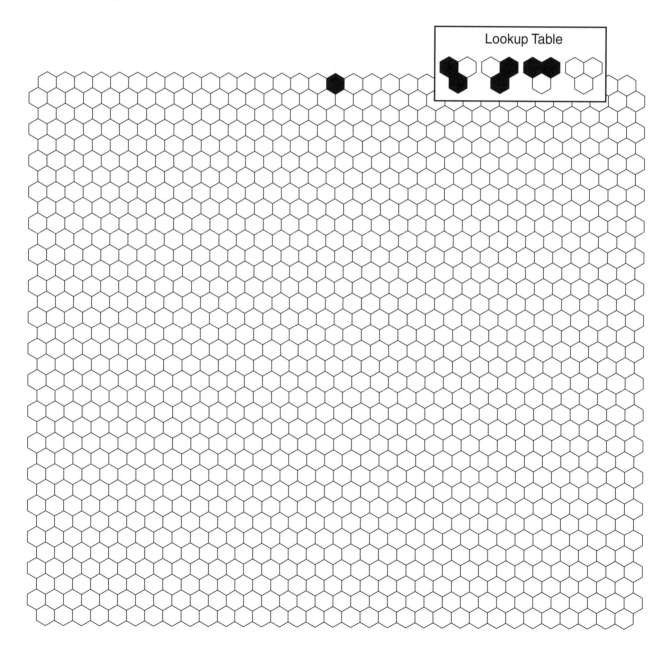

Use two copies of this sheet to do the following:

1. Create your own seed using three to five cells in the top row, then use the lookup table that you used on sheet 1. Describe your results. Compare your results with those of other students.

2. Create your own lookup table, perhaps using a different number of cells as a part of your rule. Your teacher may have suggestions. How does your model turn out? How does it compare with those of other students? Try to generalize patterns and properties that you observe by looking at other students' lookup tables, starting seeds, and final results.

A variation of CAs involves adding a little chaos to the process, by having the rule work only with a certain probability. For this model of an oil spill—as seen from the side, looking through the earth—the cells look more like bricks. Each cell in a row has two neighbors in the row below it, one to the left and one to the right. A shaded cell represents oil having spread, or percolated, past a soil particle; the particles themselves are the intersections of a cell and its neighbor below. The rule for this model is this: If a cell is shaded, a certain probability exists that it will shade each of its neighbors. Any cell could therefore have two chances of being shaded. If the cell gets shaded by its left neighbor above, no need exists to test the right one.

You will need a partner for this activity. Your teacher will assign you a probability between 0.5 and 0.75, along with instructions on how to decide whether a cell will shade each of its neighbors. Working together, you and your partner will make a model of a percolating oil spill. If your model runs off the paper, use another copy of this sheet to continue the model. How does this model differ from your previous ones? How is it similar? Compare your model with those of other students. How does yours compare with theirs? What relationship can you see between the probability used and the model's appearance?

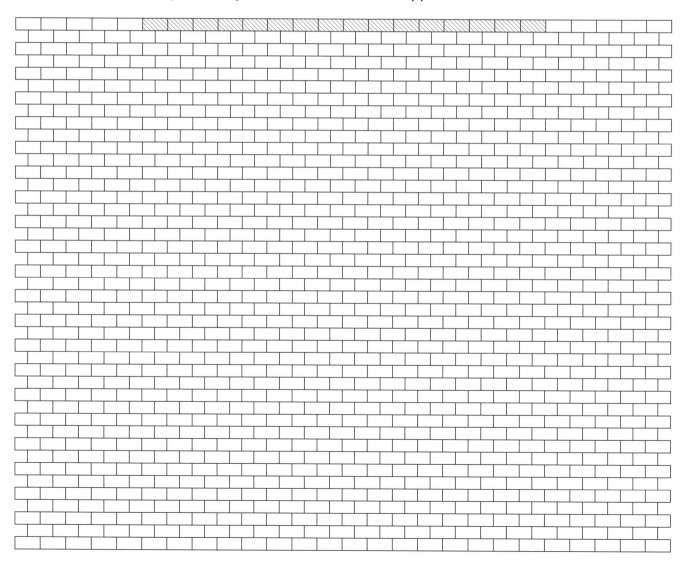

66587994545676351382706547549027176446464232957024350320324753
82034784260587057112390007281694342026466056684303822449549631
59244948684009473282734091187781126547844673719098058246314516
15055443033600350379578187659067986852415477075090158781693995
62745831149602684642100387689958689671413759698255606060467113
12876894397476172242368922366440241526703346361426215192356440
24152186841200658930516043978253683042801319927682484639912826
13637566094257066100802082250621853852075036294949888447947204
54087907676839715013527032400088978962549776529491004705067927
00225806181084730838565225644823297743524416570490362259240335
95051941955666387352226170463537828221118567423103119309425851
95705382784692373492952367939194491454888474348697060940983019
84986986974587387301959785798435974975975995959590540974975907
59503469530575389673457603246379683543235606493457430334575403
98574604523049787463534758677089702374830508542339754576755484
64035405912882178885576675836136128765570611680338878325268942
6305922822209665656198094864444090004557704378326574804121836
32464442487337229101998996434695389398249149149422045150368723
16779439154165105386134191989992683771481992487509698627494341
53012333692596111089785345540690813744860872336628971096254536
51721213483525932554406586026237395948372729118730618395115572
75840975553198680518525101356807974861184935580795959759237117
08328737265558625791321085556134713009782416233909305649610789
62402878686211005830191038766354564178192661467118739008634816
98259360615920244239211566311666119006166308047897966362253647
57838348558483747565634534231452345398459823498345089645098348
73456834568975468954678964578934568946359874539852348704538964
5367833389679543409567898899907684833829047344550885365198285
83428440652589509834517549298487828088759658007013612793723339
98868239889226057983127007416945888468601502049313309546897720
71498602456086539251282876678173535810990987653985977834568899
33309287564676798438017122337182355895359469592203401851690194
29891921843852933741554826480900123879470243137524617778989831
97198059244451529111782085039014599583566949734428459015189278

This activity allows you to try your hand at two-dimensional cellular automata. Imagine that each cell in the grid below represents a tree, with neighbors above, below, to the left, and to the right. Diagonally adjacent trees are not neighbors. If the seed for this model is a group of trees near the center of the grid, a certain probability exists that a burning tree will set each of its neighbors on fire, causing the fire to spread. Your teacher will assign your team a probability between 0.4 and 0.6 for this model. In this model, each tree can be in one of three states: unburned, burning, or dead. Once dead, a tree will stay dead and cannot set any of its neighbors on fire.

You should work in a group of three for this activity. One person should keep track of the random numbers used for spreading the fire, one should shade in red the trees that start to burn, and one should shade in black the trees that burn up and become dead. To begin, one of you should start the fire by shading a small group of cells near the center of the grid. At your teacher's direction, start at the top of the existing fire and work your way clockwise around the perimeter of the fire, testing each neighbor of a burning tree to see whether it spreads the fire. A tree will burn only for one "turn" around the fire's perimeter; once tested, a burning tree gets shaded black.

When you finish (after the fire burns out?), compare your model with those of other students. As with the oil-spill model, look for patterns relating the assigned probability to the finished model. How could you introduce the effect of wind into this model? How could you vary the density of the trees?

Nonperiodic Tilings:
The Irrational Numbers of the Tiling World

Michael Naylor January 1999

Edited by Claudia Carter, *Mississippi School for Mathematics and Science, Columbus, Mississippi*

Some shapes cover the plane only in nonperiodic fashion

Most tessellations that we see in art and nature can be reduced to one repeating unit. The honeycomb, for example, is made from one hexagon repeated in a regular pattern. A popular floor-tiling pattern might alternate octagons and squares; some manufacturers take advantage of this regularity and sell bricks in the shape of one octagon and one square connected along an edge.

Another class of tilings, however, does not share this orderly repetition. These tilings cannot be constructed with a repeating unit that is reproduced by translation alone. A cobblestone path is one example; some random arrangements of dominoes constitute another. These tilings are called *aperiodic*, or *nonperiodic*.

Because dominoes are rectangular, they can be arranged to tile the plane periodically, but some shapes will cover the plane only in a nonperiodic fashion. In 1974, Roger Penrose found a set consisting of only two tiles, called kites and darts (see fig. 1). These tiles, when arranged in accordance with simple rules, form beautiful patterns and lend themselves well to mathematical exploration. A proof that these tiles can be used to create a nonperiodic tiling of the plane but cannot be used to create a periodic tiling is beyond the scope of this article. More information about the Penrose tiles is given later in this article.

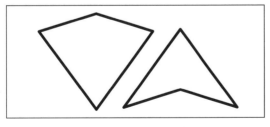

Fig. 1. A kite and a dart

This activity introduces students to nonperiodic tiling. It allows students to construct their own sets of kites and darts. Through a series of explorations, students make discoveries, form conjectures, and devise a system of notation. This activity works well either by itself or as an extension to a study of tessellations, and it can be used with a broad range of grade levels. Younger students will enjoy simply organizing shapes into interesting patterns, whereas more sophisticated students can focus on analyzing these arrangements.

TEACHER'S GUIDE

Goals: Students make conjectures, invent notation, look for patterns, and take pleasure in mathematics while learning about an unusual class of tilings.

Prerequisites: Students should have experience with tessellations.

Grade level: 7–12

Materials: Overhead transparency projector, overhead transparencies 1 and 2, one or two overhead transparencies of activity sheet 2A, an additional blank transparency, overhead transparency markers, photocopies of activity sheets, scissors, and colored markers or pencils (optional)

Preparation: Make the necessary overhead transparencies and photocopies. Carefully cut out the kites and darts from one copy of the transparency of activity sheet 2A. For additional visual effect, color the thin and thick arcs with markers of contrasting colors.

Sheet 1

Distribute copies of activity sheet 1, and read the definition of periodic tilings with the class. Emphasize that the basic repeating unit can consist of more than one shape, but it must tile the plane by translation alone. You should also point out that the pattern is pictured inside a frame; however, the frame is not part of the pattern, and students should assume that the pattern continues beyond the frame's edges.

Then have the students determine whether each pattern is periodic. If a pattern is periodic, students should find and shade the repeating unit. You might indicate that more than one answer may exist.

Michael Naylor teaches at Florida State University, Tallahassee, Florida. His interests include teacher development, geometry, technology, and recreational mathematics.

Discuss the answers as a class. You may want to show a copy of activity sheet 1 on the overhead projector during this discussion.

Dominoes. The pattern is periodic. The smallest basic unit consists of two rectangles, and six different basic units are possible. Figure 2 shows, in heavy outline, two possibilities for a basic unit that could be used to create a periodic tiling of this pattern.

Is it possible to arrange dominoes nonperiodically?

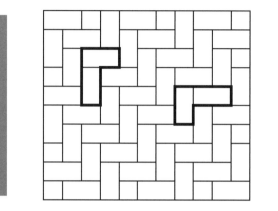

Fig. 2. Two possible basic units in the domino tiling

Patio. This pattern is a periodic tiling; the basic unit consists of a square and two rectangles. Students often decide that the basic unit is a square surrounded by four rectangles, but overlap will occur when students attempt to tile the plane.

Random rectangles. As the name suggests, no repeatable unit exists. Students may argue that the entire square region could be repeated; in fact, many flooring patterns are similar to this one. Remind the students that the square frame is not part of the pattern. Just as the digits of p never repeat, neither do these blocks; the arrangement is nonperiodic.

Scattering. This pattern appears to be highly regular, perhaps because at first glance, it seems to be simply a pattern of squares. Each square, however, consists of a triangle paired with a pentagon, and the orientation of the triangle and pentagon within these squares is random. Therefore, the pattern is nonperiodic.

Swimming pool. Many different basic units can be found in this periodic pattern. The unit should contain one large square, four small squares, and four triangles.

Starburst. This pattern is perhaps the most puzzling one of all. Students will not be able to find any portion that can tile the plane by translation alone. To help students understand why they cannot do so, point out the vertex in the center at which eight parallelograms meet. Where else will eight parallelograms meet in this manner if the pattern is continued? Nowhere. Since this vertex will never be repeated, the pattern cannot be periodic. The irrational number

$$0.123456789101112131415$$

is another example of a highly ordered pattern that does not repeat.

You might want to ask your students if it would be possible to arrange dominoes nonperiodically. One possibility is to form squares with pairs of dominoes. These squares can be oriented horizontally or vertically. If the squares are tiled with random orientations, a nonperiodic tiling similar to the "scattering" pattern will result.

Your students should now have a good grasp of what is and what is not a periodic tiling, and they should be ready to make their own sets of Penrose tiles.

Sheet 2

Explain that the class is about to explore a famous set of tiles that, when certain rules are used, can be arranged to form only nonperiodic tilings. Have the students pair up, and distribute scissors and two copies of activity sheet 2A, "Kites and Darts," to each pair. Place a kite and a dart on the overhead projector to demonstrate how the two shapes should be separated. Instruct students to cut out the shapes carefully along the straight-line segments only. To do so, first cut along the diagonal lines, creating pairs of kites and darts in a rhomboid arrangement. The curved arcs are for decoration and to help in arranging the tiles; students should not cut along any curved paths.

Ask the students whether they can arrange the shapes in a periodic tiling. Pairing kites and darts, as on the handout, is one possibility. Explain to the class that the following rule governs the arrangement of tiles and prevents the pattern from being periodic:

> Rule for placing tiles: Kites and darts must be placed edge to edge, vertex to vertex, so that thin arcs align with thin arcs and thick arcs align with thick arcs.

Penrose's original tiles did not include arcs; his rule simply specified the vertices that can be placed next to each other. The arcs were added by mathematician John Horton Conway. The designs on these tiles not only greatly facilitate placement of the tiles as specified in the original rule but also create striking patterns. The rule could be omitted entirely if the tiles were modified, perhaps with studs and notches, as shown in figure 3.

Fig. 3. A modified kite and dart

Ask students to propose some arrangements, and have the class decide whether the placements are legal. Figure 4 shows three legal placements; the tiles are placed so that the vertices and arcs are aligned. Two illegal placements are shown in figure 5; although the arcs on the two kites are properly aligned, they are not placed vertex to vertex, and although the vertices of the kite and dart are properly placed, the arcs do not meet.

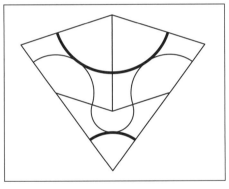

Fig. 4. Three legally placed tiles

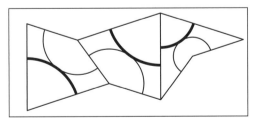

Fig. 5. Illegal placements

Allow your students a few minutes to experiment with arrangements of the tiles. You might suggest arranging tiles around one vertex at a time. Sometimes, even though tiles are arranged legally, the placements prevent a legal arrangement at an adjacent vertex. Parts of the pattern must then be dismantled and a different arrangement attempted.

Give each student a copy of activity sheet 2B. The first task should be completed as a class. The students then work in small groups on the remaining tasks.

Task 1. Kite and dart tiles may be arranged around a vertex in exactly seven ways. Find these arrangements.

Students should be able to find these arrangements in five to ten minutes. Ask students to use the overhead projector and the cut-up version of activity sheet 2A to demonstrate various arrangements until all seven have been found. Display overhead transparency 1, and have the class decide what to name each arrangement. Feel free to rotate the transparency so that students can see various orientations of the arrangements. Figure 6 shows the seven arrangements and the names that my class gave to them.

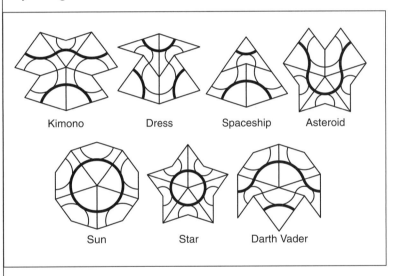

Fig. 6. The seven arrangements

Task 2. Devise a way to symbolize the arrangement of tiles around a vertex. Use your system to describe all seven arrangements.

Although nothing is wrong with sketching a picture to signify an arrangement, this process can be difficult and time-consuming. You might suggest giving each of the eight vertices a descriptive label, as in figure 7. Each arrangement can then be described by the names of the vertices of the tiles that meet at the central vertex. In this example, the seven arrangements can be designated by the following sequences of letters: kimono = PCCCC, dress = WWHH, spaceship = JCC, asteroid = PPPCC, sun = TTTTT, star = PPPPP, Darth Vader = TTWHW.

This system identifies vertices adequately, since corners or wings can be placed in only one way.

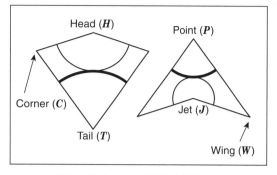

Fig. 7. One possible labeling

A more sophisticated system that differentiates between left and right corners and also between left and right wings is more powerful and allows deeper analysis of the arrangements in the following task. We can modify our previous notation to include Cl and Cr for left and right corners and Wl and Wr for left and right wings. Moving clockwise, the vertices are then more accurately designated as kimono = PCrClCrCl, dress = WrWlHH, spaceship = JClCr, asteroid = PPPCrCl, sun = TTTTT, star = PPPPP, Darth Vader = TTWlHWr.

Task 3. Now that you have symbolized the seven arrangements, look for patterns in the notation. Describe which tiles may be included at vertices with other tiles. For example, if your notation system numbers the vertices of your tiles 1–8, your rules might be similar to the following: "If two 2s meet at a vertex, then a third 2 must also meet there" or "If 4 is a vertex, then neither 5 nor 6 may be at that vertex."

This task is an excellent opportunity for students to think logically. By using the notation described previously, the following are a few of the many rules that your students may devise.

If J, then JCC (the spaceship).
P can only go with P or C.
If P, then the number of Ps must be odd and the number of Cs must be even.
P, C, and J cannot go with H, T, or W, and vice versa.
If T, then TT; if TTT, then sun.
If TW, then Darth Vader.

With the more sophisticated system, the following two rules are possible:

If ClP, then CrClP.
If PCr, then PCrCl.

Task 4. Some of the seven arrangements force the placement of other tiles around the outside of the arrangement; that is, only one way may exist to add tiles legally at certain places around the perimeter of these figures. For each of these seven arrangements, construct the entire forced region, if any. Give a name to the new shape. Hint: Use your rules!

If the students want an example, you can show them how the star forces the placement of ten kites around the perimeter. The rules that the students have written will be very helpful in this task. The spaceship and sun force no additional tiles; the dress forces two kites; Darth Vader forces at least four kites—actually eight kites and three darts, which is easier to realize with the two sophisticated rules mentioned in task 3; the kimono forces at least ten kites and six darts—actually fourteen kites and eight darts; and the star forces ten kites with pleasing radial symmetry. Students will need at least two sets of tiles to determine the region forced by the asteroid. The forced region contains thirty-three kites and seventeen darts in its immediate vicinity and forces many others further out. Constructing this asteroid field makes a good project.

Task 5. By carefully arranging the tiles, you can cause some curves on the tiles to connect to themselves to form closed loops. Find closed loops of various lengths and sizes. Make a conjecture.

Penrose and Conway proved that all closed loops have fivefold symmetry and enclose a region with fivefold symmetry (Gardner 1989, 9). Four such loops are illustrated on overhead transparency 2a.

SOME AMAZING PROPERTIES OF PENROSE TILINGS

- The Penrose kites and darts can be made by constructing a parallelogram with angles of 72 degrees and 108 degrees, dividing the long diagonal in the ratio of the golden mean, and connecting the point of division to the other two vertices (see overhead transparency 2b). The 108 degree angle will be split into angles of 36 degrees and 72 degrees. The golden mean, or golden ratio, is designated by the Greek letter phi;

$$\phi = \frac{(1 + \sqrt{5})}{2},$$

or approximately 1.618.

- The ratio of kites to darts in an infinite tiling is equal to the golden ratio. Activity sheet 2A has twenty-one kites and thirteen darts. The numbers 21 and 13 are a pair of Fibonacci numbers; their ratio is approximately the golden ratio.

- Conway proved that an infinite number of different nonperiodic tilings of the plane using only kites and darts exist (Gardner 1989, 9).

- Penrose proved that every finite portion of any nonperiodic tiling can be found in every other nonperiodic tiling. Further, any circular region of diameter d is duplicated within that same pattern no further away than the cube of the golden ratio times d, ~ 2.11d, measured from perimeter to perimeter. Therefore, any finite region of any pattern is infinitely duplicated in every possible tiling (Gardner 1989, 10).

- Every tiling is made up of large, overlapping decagons called cartwheels, as shown on overhead transparency 2c; that is, if you select any point in any tiling, you will find that it is enclosed in one or more of these cartwheels. Figure 8 shows three of these cartwheels; note that each is oriented differently from the one shown in overhead transparency 2c.

Many more startling properties of Penrose tilings exist. Martin Gardner describes them in his book Penrose Tiles to Trapdoor Ciphers. You may even wish to make from thick cardboard a set of kites and darts with colored arcs for exploration.

REFERENCE

Gardner, Martin. *Penrose Tiles to Trapdoor Ciphers.* London: W. H. Freeman & Co., 1989.

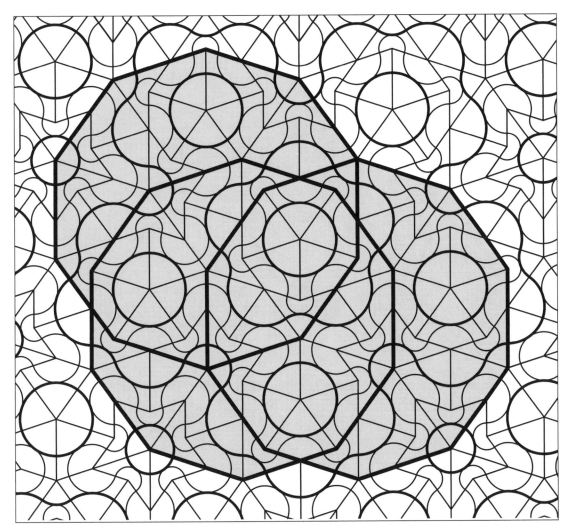

Fig. 8. Three cartwheels

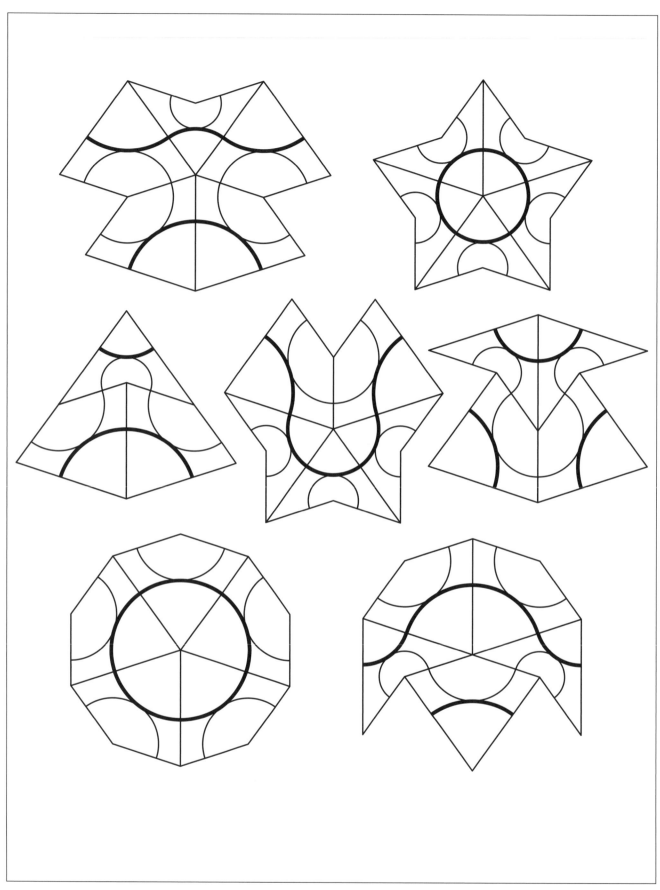

Overhead Transparency 1: Seven Arrangements

From the *Mathematics Teacher*, January 1999

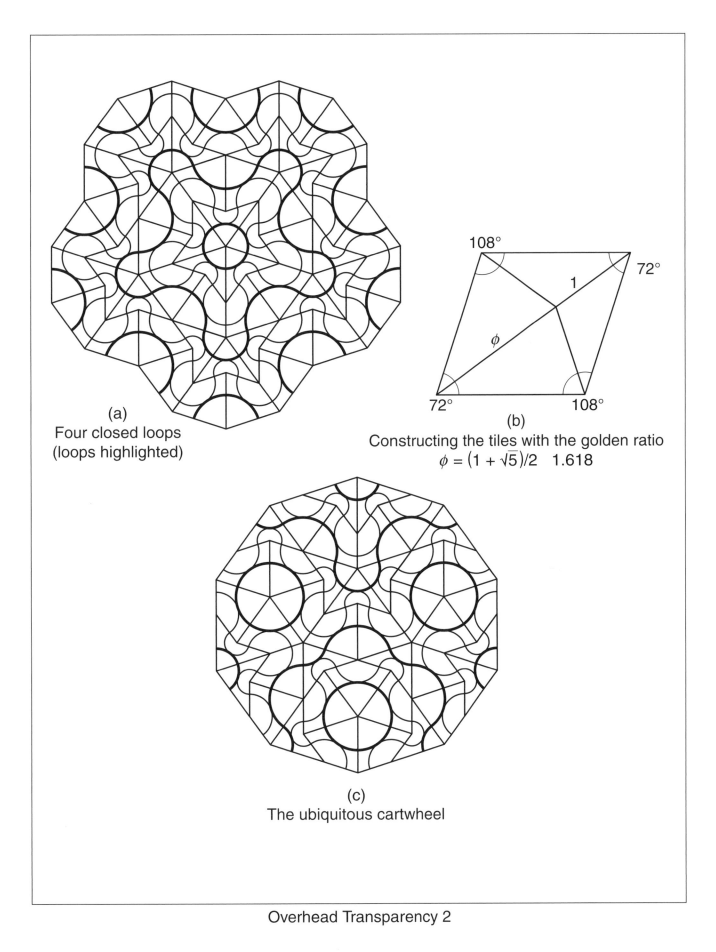

(a)
Four closed loops
(loops highlighted)

(b)
Constructing the tiles with the golden ratio
$\phi = (1 + \sqrt{5})/2 \approx 1.618$

(c)
The ubiquitous cartwheel

Overhead Transparency 2

Definition: Periodic tilings are tessellations that use a basic unit of one or more shapes that can cover the plane using translations only. Neither rotations nor reflections are allowed.

Which of the following are periodic tilings? Shade in the basic repeating unit, or write *nonperiodic*.

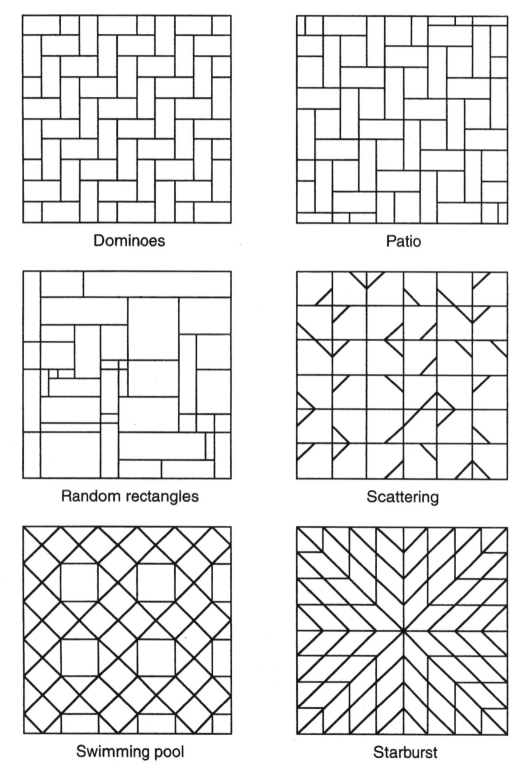

Dominoes

Patio

Random rectangles

Scattering

Swimming pool

Starburst

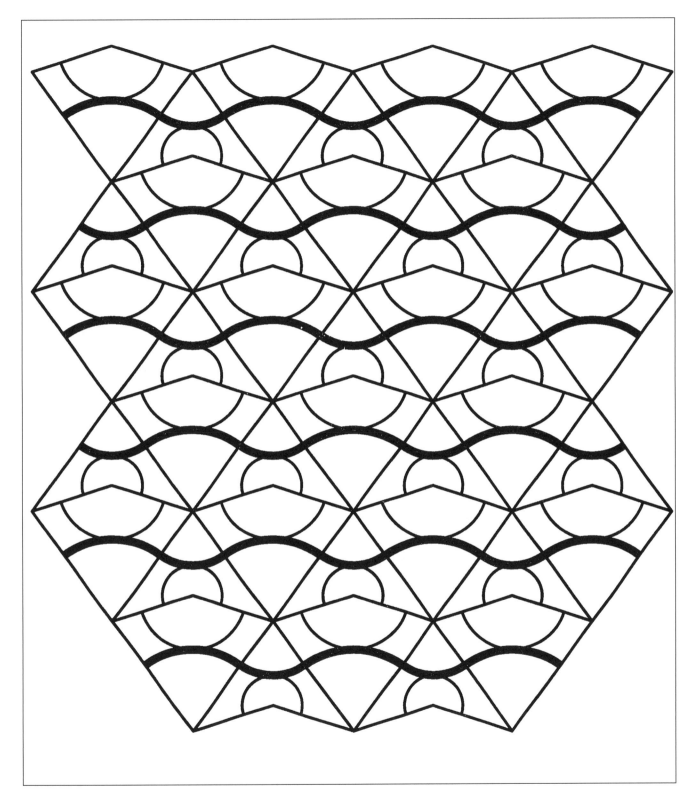

Penrose kites and darts.
Cut along straight-line segments only
(21 kites, 13 darts).

From the *Mathematics Teacher*, January 1999

Task 1: Kite and dart tiles can be arranged around a vertex in exactly seven ways. Find these arrangements.

Task 2: Devise a way to symbolize the arrangement of tiles around a vertex. Use your system to describe all seven arrangements.

Task 3: Now that you have symbolized the seven arrangements, look for patterns in the notation. Describe which tiles may be included at vertices with other tiles. For example, if your notation system numbers the vertices of your tiles 1–8, your rules might be similar to the following: "If two 2s meet at a vertex, then a third 2 must also meet there" or "If 4 is at a vertex, then neither 5 nor 6 may be at that vertex."

Task 4: Some of the seven arrangements force the placement of other tiles around the outside of the arrangement; that is, only one way may exist to add tiles legally at certain places around the perimeter of these figures. For each of these seven arrangements, construct the entire forced region, if any. Give a name to the new shape. Hint: Use your rules!

Task 5: By carefully arranging the tiles, you can cause some curves on the tiles to connect to themselves to form closed loops. Find closed loops of various lengths and sizes. Make a conjecture.

Forest Fires, Oil Spills, and Fractal Geometry: An Investigation in Two Parts

L. Charles Biehl

February 1999

Edited by Claudia Carter, *Mississippi School for Mathematics and Science, Columbus, Mississippi*

Part 2: Using Fractal Complexity to Analyze Mathematical Models

TEACHER'S GUIDE

Performing analysis on mathematically generated models of natural phenomena can provide insight into the properties of these phenomena. This article is the second of two parts. The first article, "Cellular Automata and Modeling Natural Phenomena," appeared in the November 1998 issue of the *Mathematics Teacher*. In that article, students used probability and cellular automata to generate fairly realistic models of forest-fire-burn patterns and to develop profiles of percolating oil spills. The activities in this investigation use the mathematical models of forest fires and oil spills generated in the activities in the first article. For both models, a definite connection exists between the probability used to generate the model and its physical properties. See figure 1.

A definite connection exists between the probability used to generate the model and its physical properties

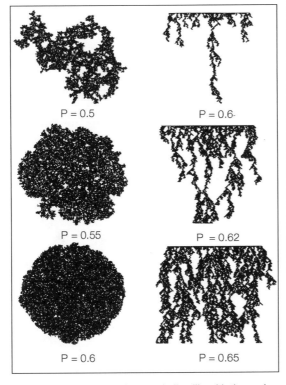

P = 0.5
P = 0.6

P = 0.55
P = 0.62

P = 0.6
P = 0.65

Fig. 1. Sample forest fires and oil spills with the probabilities used to generate them

In the oil-spill model, the outline of the percolation generally becomes more complex as the probability is raised from 0.5 through 0.6. When obvious and extensive percolation begins, the profile of the spill becomes increasingly dense. With probabilities of more than 0.6, the increase in the complexity of the model results from the "fingering" that takes place deep in the ground. As discussed in the previous article, the complexity of the profile of the spill will begin to decrease with higher probabilities, when the oil begins to percolate freely. In a similar manner, the forest-fire-burn patterns tend to be "smoother" with very low or very high probabilities; below 0.4, the fire tends to burn out quickly, and with probabilities above about 0.6, the fire spreads freely, producing a smoother boundary, nearly circular in the case of actual fires.

One way to determine how complex these figures are, regardless of their size, uses a measure called *fractal complexity*, or simply *complexity*, which measures the degree of irregularity of an object. Before determining how to find the complexity measure, students must understand what it is. Sometimes the term *fractal dimension* has been used as a synonym for fractal complexity. Euclidean shapes are referred to as having 0, 1, 2, or 3—or 4 or more?—dimensions; in a sense this number is also their measure of complexity. The boundaries of irregular objects, however, like the Sierpinski triangle or these models, are too jagged and rough to measure with integers. Even though the boundary is technically a line, the degree to which it is jagged or bent can be measured. Complexity is measured using nonintegral values; the more jagged and irregular the object is, the higher the decimal portion of the complexity measure. For a more detailed discussion, see Peitgen et al. (1992).

L. Charles Biehl teaches at The Charter School of Wilmington, Wilmington, Delaware. His interests include bringing discrete mathematics, especially twentieth-century mathematics, into the classroom.

Editorial Comment: See page 117 for Part 1 of this activity.

The measure of complexity used here on the oil-spill model is based on the Hausdorff dimension (D) of an object, also known as its complexity measure. Euclidean objects, such as line segments, rectangles, cubes, and so on, are "D-dimensional." Consider such a D-dimensional object, and suppose that the length of this object, measured in some way, is P. Further suppose that this object can be constructed from N identical objects that are similar to, but smaller than, the original, with each of these smaller objects having a length of p. For example, for $D = 3$, a cube of length $P = 2$ could be constructed of $N = 8$ smaller cubes of length $p = 1$. In general, the relationship between these quantities is expressed in the power equation

(1)
$$N = \left(\frac{P}{p}\right)^D .$$

Since D is the measure of complexity that we seek, it is necessary to solve for D, so that

(2)
$$\log(N) = \log\left(\frac{P}{p}\right)^D ,$$

(3
$$\log(N) = D \log \frac{P}{p} ,$$

(4)
$$D = \frac{\log N}{\log \frac{P}{p}} .$$

That is, the complexity, or dimension, of an object is the log of the number of smaller shapes that it takes to make a larger one of the same shape divided by the log of the scale factor between the small and larger shapes. In the example given referring to a cube, $p = 1$, $P = 2$, and $N = 8$. Substituting these values into equation (4), we get

$$D = \frac{\log 8}{\log \frac{2}{1}}$$
$$= \frac{\log 8}{\log 2}$$
$$= 3,$$

showing that a cube has a complexity of exactly 3, which is the value for D that would be expected for a cube. See figure 2a. Similarly, a larger cube of side length 3, 4, 5, . . . , n could be made out of 27, 64, 125, . . . , N^3 unit cubes, but the complexity is

The larger the decimal portion of the complexity measure, the more complex the object

still measured as 3. This same approach can be used to demonstrate that a square of side length 2, 3, 4, . . . , n has a complexity of 2, using 4, 9, 16, . . . , N^2 smaller squares. See figure 2b. The value of D can be nonintegral, depending on the values of p, P, and N; the larger the decimal portion of the complexity measure, the more complex the object.

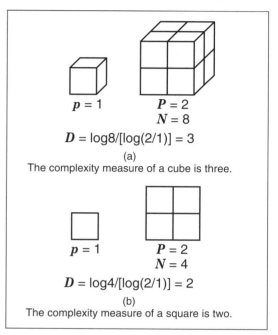

$p = 1$ $P = 2$ $N = 8$

$D = \log 8 / [\log(2/1)] = 3$

(a)
The complexity measure of a cube is three.

$p = 1$ $P = 2$ $N = 4$

$D = \log 4 / [\log(2/1)] = 2$

(b)
The complexity measure of a square is two.

Fig. 2. Computing complexity measure

Objects that are fractal exhibit three properties: they can be produced through an iterative process; they demonstrate nonintegral complexity; and they are self-similar, that is, at least parts of the object resemble the whole, meaning that the object exhibits similarity to itself on different scales. Look again at the results of sheet 1 from part 1, as shown in figure 3, to see that the figure contains smaller copies of itself.

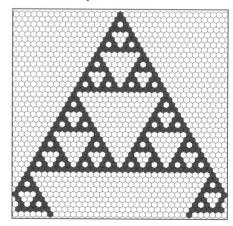

Fig. 3. The self-similar Sierpinski triangle as a cellular-automata model

Using Activities from the *Mathematics Teacher* to Support *Principles and Standards*

We can capitalize on this last property to compute the complexity of irregular shapes. Looking at the oil-spill and forest-fire models, we see evidence of a pattern of large pieces of the boundary that closely resemble smaller sections in jaggedness. Unlike with the cube in the previous example, we have no way of precisely identifying what these pieces are, let alone measuring them, because of the degree of randomness of the generation process.

For irregular objects that have not been generated using the precision of the cube or the Sierpinski triangle, a technique called a *box count* is used to collect data on the perimeter of the object to approximate the relationship between the overall perimeter of the object and the size of the unit used to determine that perimeter. In this process, an object's size, in this case its perimeter, is determined by the number of boxes on a grid containing it, using successively smaller boxes.

Imagine a large rectangle drawn on paper. Overlay the drawing with a grid of squares one inch on a side, and count the number of squares that contain part of the rectangle—not the interior of the rectangle, just its perimeter. A new count using a grid of squares one-half inch on a side should yield a count roughly twice as large. The "roughly" is the result of the count's not being exactly twice near the corners. Using grids that are successively smaller by a factor of one-half would yield counts that would be about twice as large every time that the grid size is cut in half, which means halving the length of the grid squares. Plotting these data on graph paper using ordered pairs of the form (# of boxes/inch, # of boxes containing the object) suggests, in the long run, a linear relationship with a slope showing the doubling of box counts from one size grid to the next.

Consider using a box count on the oil-spill model. Because of its irregularity, doubling the number of box sides per inch will produce a count that is more than twice the previous count. A smaller box size picks up more than twice as many of the object's nooks and crannies that would have been skipped by using a bigger box. Plotting the ordered-pair data yields a nonlinear progression of points. The data points should suggest the graph of a power equation, similar to the Hausdorff equation,

$$N = \left(\frac{P}{p}\right)^D.$$

If P is 1 inch and p is the length of the successively smaller grid boxes, then the ratio P/p equals the number of box sides per inch and N corresponds to the number of grid boxes that contain part of the perimeter of the object, yielding data points of the form $(P/p, N)$. These box-count data could then be substituted into the Hausdorff equation to approximate the value of D. However, inconsistencies in counting caused by the complexity of the perimeter make it possible for the value of D to fluctuate significantly among approximations. Some further analysis gets us past this obstacle.

Using logarithms to transform data allows students to determine whether two variables are related linearly, exponentially, or according to some power rule. A straight line is created using ordered pairs in the form (x, y) for linear data; $(x, \log y)$ for exponential data, for example, population growth; and $(\log x, \log y)$ for power rules. Looking at the Hausdorff equation as it was solved for D, notice in expression (3) that the equation

$$\log(N) = D \log \frac{P}{p}$$

bears a striking resemblance to a linear equation in the form $Y = MX$, where Y is log (N), M is D, and X is log (P/p). Plotting the logs of the box-count data should yield a straight line with a slope of D, the complexity. Plotting the raw data on log/log graph paper yields the desired results and does not require taking the logs of the data. Log/log, or *double-log*, graph paper has axes with equal intervals representing successive powers of a base. For example, the graph paper shown on sheet 2 has equal intervals between 1, 2, 4, 8, . . . , that is, powers of 2, and between 1, 10, 100, . . . , that is, powers of 10.

Grade levels: 9–12

Objectives: Through analyzing the oil-spill and forest-fire models, students use data plotting, line fitting, and slope to understand the relationships between the visual models that they have produced and the characteristics of the natural phenomena that they are modeling. The students will better understand the actual phenomena, and such understanding might eventually lead to a means of controlling these phenomena. The mathematical concept germane to this activity is an understanding of fractal complexity.

Materials: Several photocopies of oil-spill and forest-fire models generated using a variety of probabilities from "Cellular Automata and Modeling Natural Phenomena"; a set of transparencies made from grids for box counting (one division/inch, two divisions/inch, four divisions/inch,

Consider using a box count on the oil-spill model.

eight divisions/inch, and twelve divisions/inch), such as those found in *Graph Paper from Your Computer or Copier* (Craver 1996) or on such Web sites as www.rredware.com/russ18.htm or perso.easynet.fr/~philimar/graphpapeng.htm; copies of the activity sheet, copies of double-log graph paper, included as sheet 2; a sharp pencil or fine-pointed pen; tracing paper (optional); and millimeter rule

Prerequisites: The completion of the first part of this activity, "Cellular Automata and Modeling Natural Phenomena"; the ability to plot data points, "eyeball" a line of best fit, and compute the value of the slope of a line segment. Experience with the concept of logarithms may also help the student better understand the graph paper and the transformation of the data. Although the description of the activity refers specifically only to the oil-spill models, the procedure is identical for forest-fire models.

Conducting the investigation

Students can work in groups of two, three, or four, recording their progress on the activity sheet. They very carefully and accurately trace the outline of their oil spills with a sharp pencil or fine-pointed pen in one continuous line. Students can use tracing paper, or they can use the model itself. The line should not include any "holes" in the interior of the spill. Using transparencies made from the grids, students perform a box count on this outline and record their data as ordered pairs (# of boxes/inch, # of boxes containing the outline). Each team does not need a copy of all sizes of grids; they can certainly share. Teams plot these data on the double-log graph paper on sheet 2, remembering that as the numbers get larger, they get closer together. This knowledge is important for plotting approximate values; for example, on a logarithmic scale, 15 is closer to 20 than to 10.

If students have no background in logarithms, point out that the axes on log-log graph paper are designed so that as numbers get bigger, they get closer together. The numbers that are equidistant on the axes are increasing powers of the base.

Students should plot five data points: $(1, N1)$, $(2, N2)$, $(4, N3)$, $(8, N4)$, and $(12, N5)$, where N is the box count for the corresponding boxes-per-inch grid. If plotted carefully, these points should closely approximate a straight line. If an outlier or two exists, the students may need to redo that specific box count, or at the very least, check the care with which they plotted the point on the graph. After plotting the points, students should use a straight-

edge to eyeball and draw the line of best fit for the data. Remind students that this line should minimize the total distance that the data points lie from it.

To find the slope of the line as accurately as possible, extend the line to make it as long as possible. A simple way to find the slope of this line is to make it the hypotenuse of a right triangle, measure the legs in millimeters, and then calculate rise over run. This slope represents the complexity of the oil-spill outline. The "rougher" the outline, the higher the value should be, but it should not be much over 1; the "smoother" the outline, the closer the value should be to 1, since a perfectly smooth outline would be a straight line or smooth curve with a complexity measure of exactly 1.

Concluding the activity

Careful counting, plotting, and measuring should make students realize that spills generated with the lowest and highest probabilities result in the least complex outlines. Those spills in the critical range for percolation will demonstrate the most fingering and ultimate complexity. However, sometimes this notion is not borne out by the students' work. The teacher may need to emphasize that students must be precise in outlining the spill, counting the boxes, plotting the points, and finding the line of best fit. A small error in any of these processes may yield unreasonable results. Students should be aware of this need for precision before they start the activity. In addition, they may need to redo the experiment, or parts of it, to achieve results consistent with expectations. This concept is not easy for many students to accept.

Analyzing and discussing the result

The class should now have a wide variety of oil-spill outlines with their associated complexity measures. Take all the spills, display them in increasing order of the probability used to generate them, and analyze the related complexities. Visually and numerically, the relationships among the probabilities, complexities, and appearance of the spill should become evident.

Conclusions

After all student groups have completed the activity sheet, the teacher should conduct a discussion that addresses the following ideas and questions.

1. Referring specifically to forest fires, how would it be possible to compare the burn patterns of real fires with these models?

A small error in any of these processes may yield unreasonable results.

It is possible to fly over fires and photograph the burn patterns, either as the fires burn or after they are out.

2. How does the complexity of a burn pattern supply information about the severity of the fire?

Fires that are either very minor or very severe will have complexities close to 1; a minor fire burns itself out quickly and has a relatively smooth perimeter. A really severe fire burns fiercely in an almost circular pattern but has a relatively smooth perimeter. Patterns with high complexity should form the basis for further research because the complexity provides clues as to materials and conditions for the fire.

3. How does knowing the complexity of burn-pattern models help identify potential damage from a real fire?

When a fire starts, a photograph of the burn pattern could permit computing its complexity, which could then be compared with models having the same complexity, yielding information about the potential severity of the fire.

4. How does the probability used to generate a fire model relate to the actual conditions and materials involved in a real fire?

Such factors as brush, high tree density, dry conditions, and evergreen trees correlate with higher probabilities; wet conditions, certain types of trees, and time of year could contribute to a lower probability.

5. For both the forest-fire and oil-spill models, how does the range of the complexities relate to the corresponding range of probabilities?

Both models have a set of critical probabilities that produce the most complex perimeters; in addition, a low probability makes the fire or spill simpler as well as smaller, whereas low complexity relating to high probability indicates a very severe problem. Those complexities and probabilities in the critical range relate to the difficulty associated with a highly dispersed and complicated perimeter.

6. In an oil-spill model, the probability used to make the model represents the porosity of the soil. How could data relating to porosity be collected at the sites of real spills?

A set of core samples would yield this information.

7. What other kinds of materials could produce a percolation pattern similar to that of the oil-spill models?

This question is one for brainstorming and conjecturing without further research; examples include broken containers of radioactive waste, contents of derailed chemical transport trains, airplane crashes, and so on.

8. What kind of experiments would need to be performed to relate the properties of real spills to the models?

Create small spills in various kinds of soils and other ground materials, take core samples at real spills to collect percolation data, experiment with various kinds of polluting materials in soil samples.

9. How can data from the models be used to prevent real spills?

Percolation patterns can reveal how strong any preventive measures need to be to avoid a spill; knowing the potential percolation pattern in a given area could facilitate making recommendations about the kinds of containers needed for waste or even the maximum amount of waste to store at a given site.

10. What kinds of other natural phenomena exhibit the properties of fractals?

This last question can pave the way for further study. Examples include the shape of most plants, clouds, thunderstorms and lightning patterns, systems of fault lines, forests themselves, growth patterns of spreading plants, and spread of disease.

Internet Sites Featuring Cellular Automata

Digital Cat's Java Resource Center: (www.javacats.com/US). Enter "cellular automata" in the search box for interactive cellular-automata activities.

Holmes, Rich; Links to Cellular Automata Games: (web.syr.edu/~rsholmes/games/cell_aut). This site is an excellent source of links for many models and games using cellular automata.

Rucker, Rudy; CelLab: (www.mathcs.sjsu.edu/faculty/rucker/cellab.htm). Downloadable cellular-automata software gives demonstrations and examples.

Web Interactive Cellular Automata Group: (www.aridolan.com/wica). This site is under development for research in the use of cellular automata

Both models have a set of critical probabilities that produce the most complex perimeters

in many diverse applications, from Web page authoring to art.

A Java Applet That Creates the Models from These Activities

Biehl, L. Charles; Java Simulations of Forest Fires and Oil Spills: (dimacs.rutgers.edu/~biehl /camenu.html). This applet was created by Ed Young of Wilmington, Delaware, who worked to ensure a high-quality investigation activity. Comments to him from the site will be greatly appreciated.

References

Craver, John S. *Graph Paper from Your Computer or Copier*. Tucson, Ariz.: Fisher Books, 1996.

Peitgen, Heinz-Otto, Hartmut Jürgens, and Dietmar Saupe. *Introduction to Fractals and Chaos, Vol. 1, Fractals for the Classroom*. New York: Springer-Verlag, and Reston, Va.: NCTM, 1992.

Note that additional references are included in Part 1 of this activity in the November 1998 issue of the *Mathematics Teacher*.

1. Referring to the oil-spill and forest fire models that you and your classmates generated, describe the relationship between the probabilities used to make the models and the jaggedness of the outlines of the models.

2. Record the results of your box counting for one of each model in these tables:

Forest Fire			Oil Spill	
Probability used: _____			Probability used: _____	
Boxes per Inch	Box Count		Boxes per Inch	Box Count
1			1	
2			2	
4			4	
8			8	
12			12	

3. For each set of data, do the following:

   a) Plot the data as ordered pairs on the double-log graph paper.

   b) Using a straightedge, draw a straight line that best fits the data set.

   c) Make this straight line into the hypotenuse of the largest right triangle that will comfortably fit on the graph.

   d) Measure the length of each leg in millimeters, and calculate the slope of the line. Record your results.

Forest Fire	Oil Spill
Length of vertical leg (y): _____ mm	Length of vertical leg (y): _____ mm
Length of horizontal leg (x): _____ mm	Length of horizontal leg (x): _____ mm
Slope of the line (y/x)	Slope of the line (y/x)
(to the nearest thousandth): _____ mm	(to the nearest thousandth): _____ mm

4. The slopes that you found in question 3 represent the complexity measure of each model. After all groups have completed their calculations, gather data from them to complete the following tables. List probabilities in increasing order.

Forest Fire			Oil Spill	
Probability	Complexity		Probability	Complexity

5. Using the completed table for question 4, revise your answer to question 1 by describing the complexities of each kind of model as a function of the probabilities used to generate them. Use this description to contribute to the class discussion.

Box Count

Number of Boxes per Inch

Calculating Human Horsepower

Pam Cox and Linda Bridges   March 1999

Edited by Claudia Carter, *Mississippi School for Mathematics and Science, Columbus, Mississippi*

TEACHER'S GUIDE

Horsepower is a standard unit of power; it measures the rate at which work is done. Although horsepower was originally intended to be related to the rate at which a horse could deliver work, such a connection now has little meaning. Specifically, one horsepower is the rate of work required to raise a weight of 550 pounds a distance of one foot in one second of time, or approximately 746 watts. This equivalent wattage is the preferred way of expressing power in the commonly accepted Système International d'Unités, the modern form of the metric system.

This activity is a cooperative-learning activity in which students calculate the horsepower of an individual. Students work in groups to collect and analyze data. Specific objectives include solving literal equations for a particular variable, evaluating expressions when given specific values of variables, and converting from power to horsepower and from given units of weight and time to other desired units. The activity culminates in the students' application of the objectives as they compute the human horsepower of a member of the group. This opportunity to understand the relevance of solving literal equations helps students understand how mathematics is used in the real world.

Grade levels. This activity, which is suitable for either a prealgebra or algebra class, requires skills in converting units of measurement, solving equations, and evaluating mathematical expressions. Students should know such scientific terms as *force*, *work*, and *power*. Teachers should discuss the content of the opening paragraph with the class before beginning this activity.

Materials. stopwatch, bathroom scales, tape measure, calculator

Procedure

Students work in groups of four. After students are assigned to groups, each member of the group receives one of four job assignments. Teachers may wish to assign jobs to the group members so that the student who is to be weighed will not be embarrassed.

> *This activity is suitable for an algebra or prealgebra class*

The teacher should avoid drawing attention to the size of an overweight or underweight student. The jobs are as follows:

- Test subject—this student climbs the stairs and has her or his horsepower calculated.
- Statistician—this student records all measurements during the lab activity.
- Measurer—this student weighs the subject, finds the height of a set of stairs, and uses a stopwatch to record the time that the subject takes to climb the stairs.
- Secretary—this student completes the lab report with input from the group.

The students complete sheet 1 as an individual or group assignment before conducting the experiment. On the day of the experiment, all members of a group go to a designated staircase and conduct the experiment to discover the "human horsepower" of the group's subject. The measurer weighs the subject, finds the height of the staircase, and finds the time that the subject takes to climb the stairs. The statistician records all measurements and leads the group in computing the horsepower of the subject. Each member of the group should apply her or his knowledge of the objectives from sheet 1 to calculate the horsepower. When the group reaches consensus, the secretary completes sections 1–4 of sheet 2. The group then explores section 6 of sheet 2, makes conjectures, and tests its hypotheses before the secretary records its conclusion to this section of sheet 2.

Sheet 1. In this activity, students solve literal equations and then perform calculations involving these solutions. They also use the relationship that one horsepower equals 550 foot pounds per second

Pam Cox teaches at Columbus High School, Columbus, Mississippi. Her professional interests include developing activities that motivate her students to learn mathematics. Linda Bridges, teaches at Mississippi School for Mathematics and Science, Columbus, Mississippi. She is president of the Mississippi Council of Teachers of Mathematics. Her professional interests include statistics, trigonometry, and calculus.

to convert from power to horsepower. Additionally, students compute horsepower by using equations on sheet 1 and by converting units whenever necessary.

Sheet 2. Before completing sheet 2, students collect all data and compile all the information that is to be turned in for the lab report. The response to item 6 of sheet 2 should be accompanied by a convincing argument explaining why the response is true.

SOLUTIONS

Sheet 1

1. $$p = \frac{Fd}{t}$$

2. $$d = \frac{pt}{F}$$

3. $$F = \frac{pt}{d}$$

4. $$t = \frac{Fd}{p}$$

5. $$F = \frac{pt}{d} = \frac{100\,\frac{ft\,lbs}{s} \times 15s}{30\,ft.} = 50\,lbs$$

6. $$d = \frac{pt}{F} = \frac{2\,\frac{ft\,lbs}{s} \times 20s}{4.5\,lbs} = 8.89ft.$$

7. $$t = \frac{Fd}{p}$$

$$= \frac{135\,lbs \times 35\,ft.}{\dfrac{112\ ft\,lbs}{1\,s}}$$

$$= \frac{135(35)\,ft\,lbs}{1} \times \frac{1\,s}{112\ ft\,lbs} = 42.2s$$

8. $$p = \frac{Fd}{t} = \frac{175\,lbs \times 40ft}{57s} = 122.8\,\frac{ft\,lbs}{s}$$

9. Find power first, then convert to horsepower:
$$p = \frac{Fd}{t} = \frac{150\,lbs \times 8\,ft}{3\,s} = 400\,\frac{ft\,lbs}{s}$$

To convert to horsepower:
$$\frac{400\ ft\,lbs}{s} \times \frac{1hp}{550\,\dfrac{ft\,lbs}{s}} = \frac{400}{550}\,hp \approx 0.73\,hp$$

10. First convert 120 feet and 3 inches to feet:

$$3\,in. \times \frac{1\,ft}{12\,in.} = \frac{3}{12}\,ft = 0.25ft.$$

120 feet and 3 inches = 120.25 feet.
Convert one minute to 60 seconds. Then,

$$p = \frac{Fd}{t} = \frac{185\,lbs \times 120.25\,ft}{60s} \approx 370.77\,\frac{ft\,lbs}{s}$$

11.
$$370.77\,\frac{ft\,lbs}{s} \times \frac{1hp}{550\,\dfrac{ft\,lbs}{s}} = \frac{370.77}{550}\,hp$$

$$= 0.67\,hp$$

12.
$$0.5\,mile \times \frac{5280\,ft.}{1\,mile} = 2640\,ft.$$

$$2h \times \frac{60\,min}{1\,h} \times \frac{60\,s}{1\,min} = 7200\,s$$

$$p = \frac{Fd}{t} = \frac{110\,lbs \times 2640\,ft.}{7200\,s} \approx 40.33\,\frac{ft\,lbs}{s}$$

$$40.33\,\frac{ft\,lbs}{s} \times \frac{1\,hp}{550\,\dfrac{ft\,lbs}{s}} = \frac{40.33}{550}\,hp \approx 0.073\,hp$$

Sheet 2

Let F_1 equal the weight of the lighter person and F_2 equal the weight of the heavier person. Since $p = (Fd)/t$, $p_1 = (F_1 d)/t$ and $p_2 = F_2 d/t$. Because distance and time are the same for both persons but $F_1 < F_2$, the power for a lighter person is less than the power for a heavier person. Hence, the horsepower for a lighter person is less than that for a heavier person under these conditions.

BIBLIOGRAPHY

Buban, Peter, and Marshall L. Schmitt. *Understanding Electricity and Electronics.* New York: McGraw-Hill Book Co., 1962.
Settles, Gary S. Software Toolworks Multimedia Encyclopedia. Grolier Electronic Publishing Co., 1992.

w = work (ft lbs); F = force (lbs); d = distance (ft); t = time (s); p = power (ft lbs/s)

When the force applied to an object is constant, the formulas on the right can be used to calculate work and power. For gravitational forces on the earth, units of measure corresponding to the weight of the object are commonly used.

$w = Fd$ = force $\times$ distance, in foot pounds

$p = w/t$ = work/time, in foot pounds per second

1 horsepower = 550 foot pounds per second

1. Using the given formulas, write a formula for power that relates force, distance, and time. _____

2. Solve the formula for distance. _____

3. Solve the formula for force. _____

4. Solve the formula for time. _____

5. A force is applied for 15 seconds through a distance of 30 feet and at a power of 100 foot pounds per second. What is the force? _____

6. A force of 4.5 pounds is applied for 20 seconds at a power of 2 foot pounds per second. Through what distance must it be applied? _____

7. How long must a force of 135 pounds be applied through a distance of 35 feet to have power of 112 foot pounds per second? _____

8. A person who weighs 175 pounds climbs a hill that has a vertical height of 40 feet in approximately 57 seconds. How much power is used? _____

CONVERSION AND HORSEPOWER

The amount of horsepower an object has can be determined by finding its power and multiplying by the conversion factor, shown at the right.

For example, let us say that a 150-pound person climbs a mountain that has a vertical height of 1000 feet in 50 minutes. How many horsepower does this person have? See the calculations at the right.

$$\frac{1\,hp}{550\,\dfrac{ft\,lbs.}{s}}$$

$$50\,\cancel{min} \times \frac{60s}{1\,\cancel{min}} = 3000\,s$$

$$p = \frac{Fd}{t} = \frac{150\,(1000)\,ft\,lbs}{3000\,s} = \frac{50\,ft\,lbs}{s}$$

$$hp: \frac{50\,\cancel{ft\,lbs}}{1\,\cancel{s}} \times \frac{1hp}{550\,\dfrac{\cancel{ft\,lbs}}{\cancel{s}}} = \frac{50}{550}hp \approx 0.09\,hp$$

9. Suppose that a 150-pound person climbs a staircase that has a vertical height of 8 feet in 3 seconds. What is that person's horsepower? _____

10. Find the amount of power that it takes for a 185-pound person to climb a vertical distance of 120 feet and 3 inches in one minute. _____

11. Find the horsepower of the person described in question 10. _____

12. Suppose that a 110-pound mountain climber climbs a mountain with vertical height of 0.5 mile in 2 hours. What is that person's horsepower? _____

From the *Mathematics Teacher*, March 1999

Go to a designated staircase, and conduct this experiment to discover the "human horse-power" of your group's subject. Your job responsibilities are as follows: test subject—climb the stairs; statistician—record all measurements during the activity, and lead the group in calculating the "human horsepower" of the subject; measurer—weigh the subject, find the vertical height of the set of stairs, and use a stopwatch to record the time that the subject takes to climb the stairs; secretary—complete the lab report with input from the group.

1. Problem: _____

_____

2. Materials: _____

3. Procedure: _____

_____

_____

_____

_____

_____

_____

4. Measurements:

   force (weight of "subject") = _____ pounds

   distance (vertical height of staircase) = _____ feet

   time for subject to climb stairs = _____ seconds

5. Calculations of horsepower of subject (show all work):

6. Would the horsepower change if a lighter person climbed the same flight of stairs in the same amount of time? If so, in what way? Write a paragraph justifying your conclusion. How did you arrive at your conclusion? _____

_____

_____

_____

_____

Graphing for All Students

William L. Blubaugh and Kristin Emmons

Edited by Claudia Carter, *Mississippi School for Mathematics and Science, Columbus, Mississippi*

April 1999

TEACHER'S GUIDE

For years, many of us have known that an understanding of concepts from algebra can be prerequisite for many careers. Changes in society and expectations of employers have made algebraic ideas even more necessary than in the past, and more people are recognizing the significance of algebraic thinking. Over the past ten years, many documents have indicated the need to provide algebra experiences to all students (NCTM 1989; Glatzer and Choate 1992). As a result, high schools across the country are changing graduation requirements to include or expand the study of algebra. This step in the right direction creates the challenge of offering an algebra curriculum that meets the needs of all students.

One challenge in furnishing a curriculum for all students is that some students have stronger background skills than others. Although some students come into an algebra classroom with a strong sense of numbers and number operations, others are struggling with these ideas. One way to bridge, or narrow, this difference is to present intuitive experiences in a prealgebra course or at the beginning of the algebra course.

For example, the concepts of variable and function are fundamental to the study of algebra. In learning these concepts, students need to understand and work with symbolic, graphical, tabular, and verbal representational forms. Many teachers are already including pattern activities in which students generate tables and rules to fit a situation, such as the "handshake problem"; paper-folding; pattern-block trains; perimeter and area extensions; and more. Often what is lacking is a visual picture of the relationship that goes beyond tables and formulas.

A particularly challenging area for prealgebra students involves interpreting graphs and charts. This article presents activities that can be used in small-group, student-centered approach. Within these activities students are expected to —

- explain their choice of a particular graph and describe how they would label the axes,

Often what is lacking is a visual picture of the relationship

- write stories that "fit" a given graph, and
- produce graphs that describe real-world situations.

Much of the graphing that is done in classrooms involves simply plotting points that have already been given in tabular form. The real-world relationship between the variables—and hence the function concept—is often lost. Students would be better able to make these connections through experiences that area more intuitive and less contrived.

The activities in this article support students as they develop their intuitive notions of graphs. Students should be able to describe the relationships expressed in the graphs and to make generalizations of inferences on the basis of the visual image. Throughout the activities, the independent variable represents time, so that students can focus on the dependent variables.

Sheet 1

Some leading questions may help the class or individual students who do not seem to know where to start. If necessary, the teacher can begin by asking such questions as these:

1. What does the height of the graph tell you?

2. What is happening to the time as we move to the right?

3. What does the horizontal line in the graph labeled "George's Dad" mean?

William L. Blubaugh teaches at the University of Northern Colorado, Greeley, Colorado. His professional interests include the use of problem solving and technology in the mathematics classroom, the psychology of learning and teaching mathematics, teaching methodology, and the design and analysis of research in mathematics education. Kristin Emmons teaches at Del Norte School, Del Norte, Colorado. Her professional interests focus on students experiences with mathematics, including engaging students in meaningful mathematics and helping them make mathematical connections and develop intuitive ideas of mathematics.

151

4. Why does George's graph touch the horizontal axis at a point closer to the vertical axis than the graph for George's Dad?

5. How would you have to be eating to make the graph be a horizontal line?

6. What possible reasons prevent George's mom's graph from touching the vertical axis?

7. What is the significance of the x- and y-intercepts?

Students should then share their interpretations of the graph with the class. After reaching some agreement on these ideas, students begin the next activity.

Sheet 2

The first part of this activity asks students to construct a pair of graphs to represent George's popcorn-eating pattern and that of his friend Alyssa. Students need to consider in their graphs that George and Alyssa both finish eating at the same time. The second part of this sheet asks students to think about and construct a graph that represents their own eating habits.

Sheet 3

This sheet asks students to explore the meaning of a graph as the relationship between the independent variable, on the horizontal axis, and the dependent variable, on the vertical axis. Possible solutions are the following: 1-A, 2-G, 3-C, 4-F, 5-E, 6-B, and 7-D. Some other matching might make sense as well.

Sheet 4

If students have difficulty in thinking of a possible situation depicted by the two graphs, the teacher might make the following suggestions. In question 1, students can consider the speed of a car that is traveling on a short side street with stop signs at each end of the street. In question 2, students can consider a student walking home from school. The student starts walking at a constant speed, then slows down and stops for a while to talk with a friend, then starts walking again at a faster yet constant pace until arriving home. The pace is faster after the stop, possible to arrive at home on time. The students should label each axis. The teacher may wish to continue with such graphs until the students understand the increasing and decreasing nature of a function graphically.

Sheet 5

These four activities have students continue developing reasonable graphs. Again, students should label both axes appropriately and be prepared to explain their reasoning to their classmates.

In summary, we have presented a few examples of graphing explorations that can assist students during the early part of their algebra experiences. The teacher can expand these graphing ideas throughout the school year and relate them to the students' common experiences. Doing bouncing-ball activities or experimenting with Newton's laws of motion can incorporate more graphs. Such activities can improve students' interpretations of graphs. Involving students with pattern activities to generate "function" rules allows students to experience the solutions graphically.

After students have had experiences communicating through general graphs, they will be much better prepared to make intelligent inferences from graphs. Students will be better able to interpret graphs and understand the relationship between the variables that the graph illustrates. The activities presented in this article should help all students build a conceptual basis for the graphs that they will use in other courses, help them interpret real-world graphs, and prepare them for more formal algebraic experiences.

BIBLIOGRAPHY

Forester, Paul A. *Algebra and Trigonometry: Functions and Applications.* Menlo Park, Calif., Addison-Wesley Publishing Co., 1994.

Glatzer, David J., and Stuart A. Choate. *Algebra for Everyone: In-Service Handbook.* Edited by Albert P. Schulte. Reston, Va.: National Council of Teachers of Mathematics, 1992.

National Council of Teachers of Mathematics (NCTM). *Curriculum and Evaluation Standards for School Mathematics.* Reston, Va.: NCTM, 1989.

———. *Professional Standards for Teaching Mathematics.* Reston, Va.: NCTM, 1991.

———. *Assessment Standards for School Mathematics.* Reston, Va.: NCTM, 1995.

George and his family were watching a movie and eating popcorn. Each family member had a bowl with the same amount of popcorn. The graph below all show the amount of popcorn remaining in the person's bowl over a period of time. Under each graph, write a few sentences describing what may have happened.

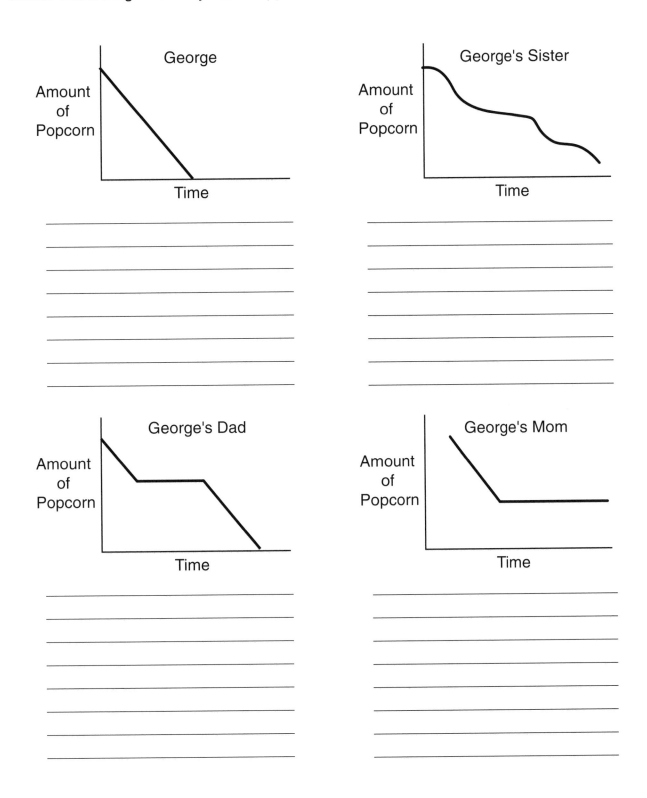

George and his friend Alyssa went to the movies, where they each brought a medium tub of popcorn. George was quite hungry and quickly ate half his popcorn, paused for a moment, and then continued eating at the same rate as before. Alyssa waited for a few minutes before she began and then ate at a steady rate. When she noticed that George had almost finished his popcorn, she gave him some of her own. They then both continued eating and finished at the same time.

1. On the axes below, show how the amount of popcorn remaining in George's tub varied over time.

2. On the same axes but using a different color, show how the amount of popcorn remaining in Alyssa's tub varied over time.

3. Examine each graph. How are the two graphs similar, and how they are different?

_____

_____

_____

4. How does your graph show that Alyssa waited for a few minutes before she began eating?

_____

5. How did you show on your graph that she gave him some of her own popcorn?

_____

_____

6. How does your graph show that they both continued eating and finished at the same time?

_____

_____

7. How does your hunger change from morning until night? On the given axes, show how your hunger changes from the time you wake up in the morning until you go to bed at night

8. Write a paragraph explaining what your graph reveals about your eating habits.

Match each of the following seven scenarios with the most appropriate graph given. As you look at each graph from left to right, remember that time is advancing.

_____ 1. We rode the roller coaster steadily to the top, then went faster and faster as we went down the other side. The speed of the roller coaster is the dependent variable of the graph, that is, the variable on the vertical axis.

_____ 2. The kettle heats before the corn begins to pop. The corn starts to pop and continues popping until almost all the corn had popped. The amount of unpopped corn in the kettle is the dependent variable.

_____ 3. A balloon was blown up in class and then let go. It flew around the room. The amount of air in the balloon is the dependent variable.

_____ 4. At the beginning of spring, the grass grew slowly and I seldom had to mow the lawn. By midsummer it was really growing, so I mowed twice a week. In fall, I only mow once in a while. The number of lawn mowings to date is the dependent variable.

_____ 5. I turned the oven on. When it was hot, I put in the cake. The cake baked for about thirty minutes. I turned the oven off and removed the cake. The oven temperature is the dependent variable.

_____ 6. We bought a pair of rabbits last year. They have had several litters, and we have so many rabbits that the pens are full. If more are born, we will have to give some away or find room for the new ones. The number of rabbits is the dependent variable.

_____ 7. I put water in the ice-cube tray and placed it in the freezer. The temperature of the water in the ice-cube tray is the dependent variable.

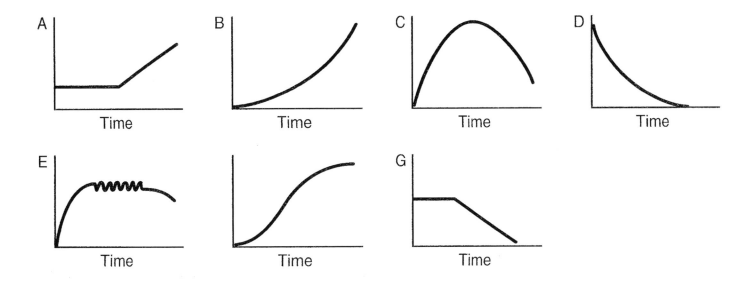

1. Think of a real-life situation that could be represented by the graph on the right. Write a story about the situation, and be prepared to read your story to the class. Be sure to label the axes of the graph.

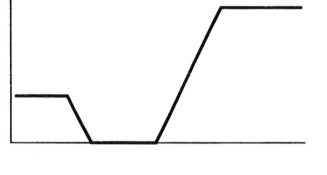

_____

_____

_____

_____

_____

_____

_____

_____

_____

_____

2. Think of a real-life situation that could be represented by the graph on the right. Write a story about it, and be prepared to read the story to your classmates. Be sure to label the axes of the graph.

_____

_____

_____

_____

_____

_____

_____

_____

_____

_____

APPLICATIONS SHEET 5

Before graphing each relationship, label both axes appropriately, and be ready to explain to your classmates the reasoning behind each of your graphs.

1. You turn on the hot-water faucet. The temperature of the running water depends on the number of seconds since you turned on the faucet.

2. As you play with a yo-yo, the yo-yo's distance from the floor depends on the number of seconds that you have passed since you started.

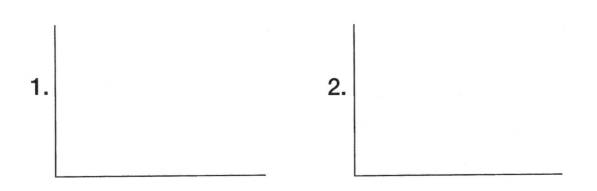

3. You go from sunlight into a dark room. The diameter of your pupils depends on the length of time that you have been in the room.

4. You pour some cold water from the refrigerator into a glass and leave it on the counter. As the glass sits on the counter, the water's temperature depends on the number of minutes that have passed since you poured it.

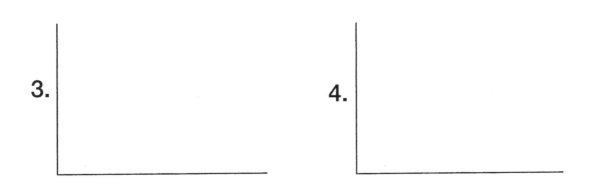

Whelk-come to Mathematics

Brian A. Keller and Heather A. Thompson

September 1999

Edited by A. Darien Lauten, *Rivier College, Nashua, New Hampshire*

How many times must a whelk be dropped to break open its shell?

Do you ever wonder why animals behave as they do? For instance, scientists have noticed that northwestern crows drop whelks, large marine snails, consistently from a height of 5 meters. Why 5 meters? The answer to this question may seem obvious; they wish to break open the shells to eat the soft meat inside. But why 5 meters? This situation raises many interesting questions that are open to mathematical investigations. For instance, how many times must a whelk be dropped to break open its shell? What factors influence the height from which the whelk is dropped? Is 5 meters the optimum height to minimize the work?

Students explore these questions through this hands-on modeling experiment. In the activity, hyperbolic functions, asymptotes, data analysis, applied statistics, and optimization problems arise naturally. Students reason about the context to understand hyperbolic functions. Finally, directions and activity sheets allow students to conduct an experiment that simulates the behavior of the crows and parallels the analysis done by Reto Zach, who investigated possible reasons for the crows' behavior.

Conjecture

Zoologist Reto Zach from the University of British Columbia studied five pairs of northwestern crows as they gathered whelks, dropped them, and ate them (1978, 1979). He hypothesized that the crows' behavior was an example of optimal foraging. Optimal foraging implies that an animal demonstrates sophisticated behavior in making decisions about its food supply. Zach conjectured that crows optimize the work done in dropping whelks. To verify or disprove his theory, Zach observed the crows' choice of whelks, their flight pattern while dropping the whelks, the height from which the whelks were dropped, and the number of times that each whelk was dropped before the shell broke.

Zach used the information that he gathered to calculate the work that crows put forth in breaking open the shells. He decided that if the crows drop the whelks from the height that minimizes the amount of work, then his conjecture would be

proved. The amount of work is directly proportional to the number of drops, the height of the drop, and the weight of the whelk. Zach observed that the crows select only the largest whelks, those that are approximately four centimeters long and that weigh approximately eight grams. Thus, an equation for the work exerted by the crows in dropping a whelk from the same height until breakage occurs is $W \approx N \cdot H$. Here, the work, W, depends on the product of the number of drops, N, and the height of the drop, H. The number of drops needed to break a shell should be a function of the height of the drop. Therefore, a secondary goal is to determine the number of drops to break open a shell when the height is given.

Experiment

To find the relationship between the height and the number of drops needed, Zach dropped whelks that he classified as small, medium, or large in length and weight. To explain the crows' selection of large whelks, Zach explored the possibility that large whelks break more readily than small- or medium-sized whelks. He dropped twelve whelks of each size from heights of 2, 3, 4, 5, 6, 7, 8, 10, and 15 meters. Each whelk was dropped until the shell broke open. He recorded the number of drops required to break each shell. Zach analyzed the data by finding a mathematical relationship between the number of drops and the height of the drop to formulate a work equation. In this activity, students use peanuts to simulate Zach's experiment.

Analysis

To analyze the data, Zach followed a series of steps that students repeat as part of this activity. Zach conjectured a model for the data, rewrote the conjectured model to express a linear relationship, used

Brian A. Keller and Heather A. Thompson, teach at Iowa State University, Ames, Iowa. Keller is exploring the effects of context-based learning with technology on students' development of symbol sense, critical thinking, and metacognitive skills. Thompson is investigating students' inquiries during calculus laboratories.

the model to transform the data, and graphed the transformed data to determine whether a line resulted. When a linear relationship resulted, he could find the equation of the line and solve for the desired variable. Thus, to find a meaningful mathematical relationship, conjectures about the form of the relationship between the number of drops, N, and the height of the drop, H, are necessary.

One assumption is that the graph of this relationship has a horizontal asymptote at $N = 1$. This asymptote exists because at least one drop is needed for any height and because as the height increases, the number of drops needed decreases toward 1.

Another reasonable, although more complicated, assumption is that the graph also has a vertical asymptote. If a whelk is repeatedly dropped from a height of 1 mm, the shell may never break. The momentum is not large enough to cause breakage. Thus, the model should account for a vertical asymptote at $H = H_{min}$ where $H_{min} \geq 0$. One strength of this activity is that students can use the context to reason and to make sense of these asymptotes.

Assuming both the vertical and horizontal asymptotes, a reasonable model for the number of drops, N, is a hyperbolic function:

Students make sense of asymptotes

$$N = 1 + \frac{k}{H - H_{min}},$$

where H is the height of the drop, H_{min} is the minimum height, and k is a stretch factor. The graph of Zach's large-whelk data, shown in figure 1, confirms that the assumptions behind the conjectured model are reasonable. The equation that uses the hyperbolic model can be transformed to give a linear relationship between $(N - 1)^{-1}$ and H, namely,

$$(N - 1)^{-1} = \frac{H - H_{min}}{k}$$
$$= m(H - H_{min}).$$

Of crucial importance is that the data can be transformed in a similar manner to give a graph of $(N - 1)^{-1}$ versus H. This graph confirms an approximately linear relationship. Students can then find the equation of the line that best fits the transformed data and can solve the resulting equation for N in terms of H. Figure 2 is based on Zach's data and shows a graph of the transformed data and the approximating linear function, along with the graph of the original data and the equation for N that comes from the linear equation.

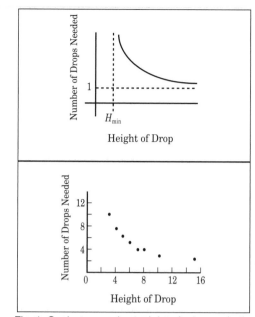

Fig. 1. Conjecture and actual data for large whelks

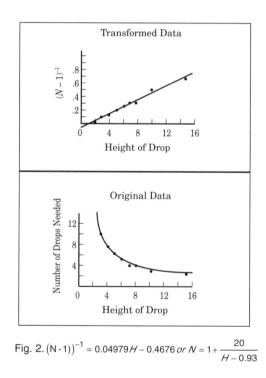

Fig. 2. $(N-1)^{-1} = 0.04979H - 0.4676 \; or \; N = 1 + \dfrac{20}{H - 0.93}$

Conclusion

After determining the equation for N, students can calculate the work done by using the general equation

$$W = N \cdot H$$
$$= \left(1 + \frac{k}{H - H_{min}}\right) H.$$

By using the specific expression for N given in figure 2, that is,

$$N = 1 + \frac{20}{H - 0.93},$$

this work equation becomes

$$W = \left[1 + \frac{20}{H - 0.93} \right] H$$

The graph of this equation, shown in figure 3, is a U-shaped curve that has a minimum at $H = 5.24$ meters. Zach observed that the crows drop whelks from an average height of 5.23 meters. The small difference between these two values suggests that the northwestern crows demonstrate optimal foraging, as Zach conjectured.

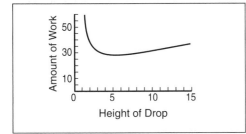

TEACHER'S GUIDE

Without leaving the classroom, students can simulate Zach's investigation of the behavior of northwestern crows. After an initial discussion of the context, students can generate, collect, and analyze data in a similar manner using peanuts in place of whelks. Although many teachers struggle with teaching rational functions (Smith 1998), the rich context illustrated in the activity sheets helps students make sense of asymptotes as they reason why the crows act as they do. The described activity can be extended to discuss oblique asymptotes, items from elementary statistics, and optimization using calculus.

Prerequisites: Students should have some exposure to functions with asymptotes. This activity can follow a brief introduction of the basic hyperbola, $y = 1/x$, and can lead to a more complete discussion of hyperbolas. This activity can be used to introduce and apply methods of translations, including vertical shifts, horizontal shifts, and stretching or compressing functions.

Grade levels: second-year algebra, advanced algebra, precalculus, statistics, and calculus

Materials: Each student needs activity sheets and a graphing calculator. Each group of three to four students needs whole, blanched peanuts that have been removed from their shells and a meter-

stick. Different kinds of peanuts produce different ranges of data. The instructor must experiment before class to determine an appropriate range over which to drop the peanuts or must structure the class to design this component into the experimental process. The data given in the answers are based on a specific brand of oily blanched peanuts. Other brands of blanched peanuts broke relatively easily, and a smaller range of 5 to 30 centimeters was needed. A brand of Spanish peanuts that broke less easily required a range of 0.5 meters to 3 meters. A sixteen-ounce bag of peanuts is usually sufficient for twelve students if the students work in groups of three to four. Some students are highly allergic to peanuts and peanut oil. Check with the students before conducting the experiment.

Directions

Sheet 1. Sheet 1 gives the context and the introduction for the investigation. Instead of distributing a copy of this sheet to each student, the teacher can use an overhead transparency of sheet 1. Before answering the questions, students can view a brief film that shows the crows' behavior. The videotape *Birds of the Lands of Four Seasons,* by Churchill Films, contains a clip about the crows. A short movie is available at www.math.iastate.edu/keller. Either of these video clips helps introduce the activity.

Sheet 1 helps students reason about the context and make conjectures to suggest a model. Students' discussion about the crows' flight paths and factors influencing the height leads to the conjecture that crows minimize the work exerted to break open the shells. Students should propose how they might conduct an experiment that is similar to the natural phenomenon. This activity helps increase students' ownership of the problem and solution. For that reason, distribute sheets 2–5 after students have discussed sheet 1. In discussing the context, students readily agree that the relationship between the number of drops and the height of the drop is needed. The students' conjecture of this relationship is essential to the analysis following data collection.

Students must reason about the need for horizontal and vertical asymptotes before sketching a graph in question 5. To discuss this need, students may require such prompts as "Is a minimum number of drops needed to break a whelk's shell?" and "Is a minimum height needed to break a whelk's shell?" Students easily understand that the minimum number of drops relates to a horizontal asymptote at $N = 1$. The need for a minimum height and vertical asymptote remains unanswered until the analysis is complete and should be discussed again at the end of the activity. After making these conjectures, students typically sketch a graph of a hyperbolic function justifying that a lesser height requires more drops and that a greater height requires fewer drops to break open the shell.

Sheet 2. Sheet 2 describes the process that groups of three or four students use to gather data. Alternatively, a sample data set could be given to the students. This activity uses peanuts in place of whelks. The peanuts are repeatedly dropped from the same height until they break. For each of the eight heights, eight peanuts should be dropped and the mean number of drops calculated. Another way to proceed is to have the class design this component of the experiment. Students should determine whether the patterns found in the heights and in the average of the number of drops support or refute their conjectures on sheet 1. At this time you may want to discuss general steps for determining a model for situations:

1. Gather the data.
2. Conjecture a model for the data using the context and the graph.
3. Rewrite the model in a linear form.
4. Transform the data, and determine whether a linear relationship results.
5. If the relationship is linear, find the equation of the line and solve for the desired variable.
6. Verify the model by comparing the final equation with the original data.

Sheets 3 and 4. Sheets 3 and 4 direct the students in analyzing the data and determining equations for the number of drops and the amount of work. The discussion on the existence and placement of asymptotes continues with questions 1(*b*) and 1(*c*) on sheet 3. Again, students may not be convinced that a vertical asymptote exists until they complete the analysis. If students recommend an exponential function over a hyperbolic function in 1(*d*), ask them questions that help them see the inconsistencies between what they have seen and their equation. For instance, you might ask about the vertical asymptote, the *y*-intercept, or the nonzero horizontal asymptote.

Question 2 emphasizes steps 3 through 6 of the process above for determining a model. Students should rewrite the suggested model in question 1(*d*) to obtain a linear relationship between $(N - 1)^{-1}$ and H and to determine an appropriate transformation of the data. After completing the table in question 2(*b*), students can use linear regression or algebra to find the equation for the line. Students tend to write their equation in terms of *x* and *y*. You may need to remind students to use H and $(N - 1)^{-1}$, respectively, which serves as a reminder to solve for *N*. From the equation for *N*, students can determine the location of the vertical asymptote. The accuracy of the equation and of the vertical asymptote can be determined by comparing the data and the equation numerically or graphically. At this point, the need for the vertical asymptote is obvious because students see how appropriate and reasonable the hyperbolic model is.

In sheet 4 students investigate the equation for the amount of work as a function of the number of drops and the height of the drop. In question 1, students write the equation for work using the equation for the number of drops from sheet 3, question 2(*d*). After graphing their work equation in question 2, students can find the height that minimizes the work. Students in second-year algebra, advanced algebra, or precalculus can use the functions on their graphing calculators to locate the minimum. Calculus students can apply the product rule and quotient rule in determining the first and second derivatives of the work function to find and verify the height that minimizes the work.

Sheet 5. On sheet 5 students repeat the analysis process for the data gathered by Zach. The goal is to determine the height for the minimum work in dropping a whelk. This task brings the activity back to the original question. The graph of the number of drops needed as a function of the height of the drop is given for large, medium, and small whelks. In distinguishing the differences among the curves, students should observe that smaller whelks require a larger minimum height. To explain this observation, students may reason that a smaller whelk may have a stronger shell or may have a smaller momentum when dropped, thus requiring a greater height.

The data that students read from the graph for the large whelks will vary. These variations will produce different equations. Also influencing the equation is the placement of the horizontal asymptote. Although the location is unclear, a horizontal asymptote at $N = 1$ seems reasonable. As students explore the work function and find the height that minimizes the work, answers should range from 5 to 7 meters. Values between 7 and 8 meters are possible but could indicate insufficient accuracy in reading the graph. By the end of sheet 5, students should have verified Zach's conjecture that crows drop whelks in a manner that optimizes work. At this point it is important to discuss the mathematics in terms of the context.

Rather than have your students complete all the activity sheets, you could have students complete only sheet 1 and sheet 5. Students can explore and discover different methods to verify or disprove Zach's conjecture. Students could discuss their methods with the class and could compare their methods with those that Zach used to support his conjecture.

Assessment

Sheet 5 may be used to assess the students individually, since students repeat the analysis process. In addition, students may write a summary of the mathematics used, the claims and justifications made, and the application of the related skills to other mathematical situations. As students reflect on the mathematics, the algorithms applied, and the sense-making approach, mathematical insight deepens.

Extensions

Rational functions with oblique asymptotes can be studied as an extension to this activity. The graph of the amount of work has oblique asymptotes. For very great heights, the curve closely resembles the graph of $y = H + k$. This equation reflects the observation that for large heights the work is essentially the energy needed to drop the whelk once. For heights close to the minimum height, the graph of the work function resembles the graph of

$$y = \frac{H_{min}k}{H - H_{min}}.$$

This equation is similar to the equation for the number of drops. Thus, for low heights the work is predominantly determined by the number of drops. Students can graph the two equations

$$y = H + k$$

and

$$y = \frac{H_{min}k}{H - H_{min}}$$

on the same set of axes as the graph of the work function to see that these two equations approximate the situation for the extreme examples of very great heights and of heights near the minimum, respectively.

In addition, students may show that the equation for the amount of work is the sum of two functions,

$$W = \left(H + k\right) + \frac{H_{min}k}{H - H_{min}}$$

This activity gives students practice in manipulating and making sense of symbolic expressions.

Additional extensions to this activity may result as students investigate the behaviors of other animals. Students may wish to search for other animals whose behaviors show optimal foraging. This activity is a natural connection between the biology or animal-ecology curriculum and mathematics.

Modifications for statistics

This activity could be also be used in a statistics class. After gathering the data, students can calculate the mean and standard deviation of the number of drops at each height. Using the standard deviations, students discuss whether the mean appropriately represents the data. If not, the median or mode could be used in place of the mean in the activity sheets. Groups of students could use the mean, median, or mode to complete the analysis, then determine which statistic produces the best results. The actual model is not likely to pass through the lower heights. Examining confidence intervals around each mean furnishes additional evidence of the appropriateness of the model. Here students can see the importance of the accuracy of a statistic representing the entire data set.

ANSWERS

Sheet 1

1. Zach observed that the crows use the second flight path, *B*, most often. One conjecture is that the path permits them to see where the whelk lands even though they lose some of the height of the drop.

2. Some factors might include the work involved in lifting the whelk, the likelihood that the shell will break, the need to minimize the chance that other birds will steal the whelk, and better aim at lower heights.

3–4. Answers will vary. Whole, blanched, shelled peanuts can represent whelks and can be dropped from differing heights to determine the number of drops needed before each peanut breaks, as well as the work involved to break it.

5. A sketch of a decreasing graph should include a horizontal asymptote at $N = 1$ and a vertical asymptote at some unknown height $H = H_{min}$, where N is the number of drops and H is the height of the drop. Certainly the

The amount of work is the sum of two functions

whelk must be dropped at least once, accounting for a horizontal asymptote. One possibility is that a whelk may never break. Repeatedly dropping a whelk 1 mm may never cause breakage. Thus, a height may exist at which insufficient force is generated to ever break open the shell. A worst-case scenario is that if a whelk is "dropped" from a 0 height, then an infinite number of nondrops will not break open the shell. This result suggests a vertical asymptote.

Sheet 2

1. Figure 4 shows the mean number of drops for each height for the data obtained by one class.

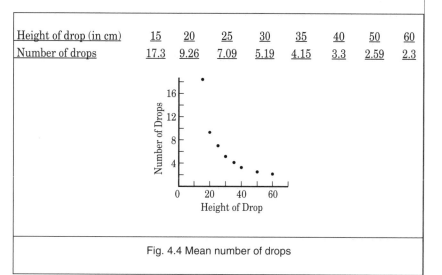

Height of drop (in cm)	15	20	25	30	35	40	50	60
Number of drops	17.3	9.26	7.09	5.19	4.15	3.3	2.59	2.3

Fig. 4.4 Mean number of drops

Sheet 3

1. *a)* The graph resembles the graph of the hyperbolic function y = 1/x shifted up and to the right. Thus, a possible equation relating the two variables is

$$N = a + \frac{1}{H - b},$$

where a is the amount that the graph is shifted up and b is the amount shifted to the right.

b) At least one drop is required to break a peanut. As the height increases, the number of drops needed approaches 1. The equation modeling the peanut data has a horizontal asymptote at N = 1.

c) Responses will vary. Some peanuts required many drops to break at 15 cm, but the existence of a minimum height is difficult to determine. Students must use caution in asserting the existence of a vertical asymptote.

d) The data could be modeled by a hyperbolic equation of the form

$$N = 1 + \frac{k}{H - H_{min}},$$

since it has a horizontal asymptote at N = 1 and a vertical asymptote at $H = H_{min}$ for some minimum height H_{min}.

2. *a)* Rewriting this equation yields the relationship

$$(N-1)^{-1} = \frac{(H - H_{min})}{k}$$
$$= m(H - H_{min})$$
$$= m \bullet H + b,$$

where $m = 1/k$ and $b = H_{min}/k$. The expression $m \bullet H + b$ is a linear expression in H.

b) Table 1 uses the sample data from question 1, sheet 2.

c) The regression equation for the relationship between H and $(N - 1)^{-1}$ is

$$(N - 1)-1 = 0.0164H - 0.223.$$

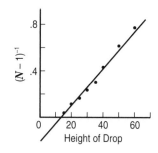

Table 1							
Transforming the Data							
Height of drop (in cm)	15	20	25	30	35	40	50	60
Number of drops	17.3	9.26	7.09	5.19	4.15	3.3	2.59	2.3
$(N - 1)^{-1}$	0.061	0.121	0.164	0.239	0.317	0.435	0.629	0.769

d)
$$N = 1 + \cfrac{1}{0.0164H - 0.223}$$

or

$$N = 1 + \cfrac{60.9}{H - 13.6}$$

The equation fits the data well except for the lowest height of 15 cm.

e) The equation

$$N = 1 + \cfrac{60.9}{h - 13.6}$$

has a vertical asymptote at $H = 13.6$ cm. Intuitively this value seems high. Dropping a few peanuts at 10 cm will confirm that peanuts can break at a height below 13.6 cm. However, for the range in question, the equation appears to model the data reasonably well.

Sheet 4

1. $$W = \left(1 + \cfrac{20}{H - 0.93}\right)H \,.$$

2.

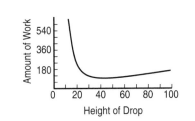

Height of Drop

3. Dropping the peanut from a height of 42.4 cm minimizes the amount of work.

4. Students may discuss their initial conjectures for the experiment, rational equations, asymptotes in the equations and what the asymptotes represent in the context, the process of finding a linear relationship with the transformed data, translations of functions, or optimizing the work to drop a peanut.

Sheet 5

1. Answers may vary. Zach wanted to investigate whether the larger whelks broke more readily and yielded more meat for the work needed to drop them.

2. The smaller the whelk, the greater the minimum height from which it must be dropped and the greater the minimum number of times that it must be dropped before its shell breaks. Possible explanations include that a small whelk will have smaller momentum than a larger whelk and that a small whelk may be structurally stronger than a large whelk.

3. Responses may vary. The data for the large whelk are given in table 2. By using the methods that we used previously, we get the equation for work to be

The model depends on the horizontal asymptote chosen for the number of drops. The actual asymptote is unclear, but a horizontal asymptote at 1 seems plausible. Using that asymptote, we get the equation

$$(N - 1)^{-1} = 0.04979H - 0.04676,$$

or

$$W = \left(1 + \cfrac{20}{H - 0.93}\right)H \,.$$

4. The graph of the work function has a local minimum at approximately (5.24, 29.56). The work does not grow rapidly beyond that point but is very large for low heights. The minimum value depends on the data that are read from the graph. Values between 7 and 8 meters are possible but may be the result of insufficient accuracy in reading the data from the graph.

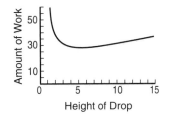

Height of Drop

5. Given the shape of the graph, the work between 4 and 6 meters is close to the minimum amount of work. One answer is to drop from 4 meters, producing less chance that

Table 2								
Data for Large Whelk								
Height of drop (in cm)	2	3	4	5	6	7	8	10	15
Number of drops	55	10	7.5	6	5	4	4	3	2.5

other birds will steal the meat. The greater height may give a better viewpoint for selecting a location for dropping the whelk. Other justifications are also possible.

6. The data from the graph and the fitted models seem to indicate that the northwestern crows drop whelks from a height that minimizes their work to break open the shell. This result supports Zach's conjecture that dropping whelks from a height of 5 meters is an example of optimal foraging.

REFERENCES

Boswall, Jeffrey. *Birds of the Lands of Four Seasons.* Churchill Films, 1987. Videotape.

Smith, Cynthia. "A Discourse on Discourse: Wrestling with Teaching Rational Equations." *Mathematics Teacher* 91 (December 1998): 749–53.

Zach, Reto. "Selection and Dropping of Whelks by Northwestern Crows." *Behavior* 67, parts 1–2 (November 1978): 134–47.

———. "Shell Dropping: Decision-Making and Optimal Foraging in Northwestern Crows." *Behavior* 68, parts 1–2 (July 1979): 106–17.

This activity is an adaptation of material from Polynomial and Rational Functions, course 4, unit 5A, Core-Plus Mathematics project. In preparing this material, the authors gratefully acknowledge the collaboration with Jim Cornette and Gail Johnston at Iowa State University and Chris Russell at Marshalltown Community College. This work was supported in part by the National Science Foundation, grant numbers DUE 9354437 and ESI 9617183. The opinions expressed are those of the authors and not necessarily those of the Foundation.

Seagulls and crows feed on various types of mollusks by lifting them into the air and dropping them onto a rock to break open their shells. Biologists have observed that northwestern crows consistently drop a whelk, a type of marine mollusk, from a mean height of about 5 meters. The crows appear to be selective; they pick up only large-sized whelks. They are also persistent. For instance, one crow was observed to drop a single whelk twenty times. Scientists have suggested that this behavior is an example of decision making in optimal foraging.

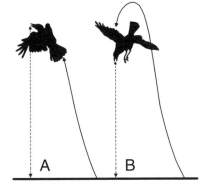

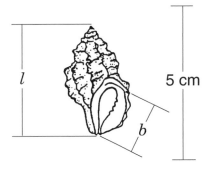

Possible Flight Paths   Large Whelk

Why do crows consistently fly to a height of about 5 meters
before dropping a whelk onto the rocks below?

Making a conjecture: Consider northwestern crows dropping large whelks.

1. Biologists observed that the crows most often drop a whelk after flying in one of the two flight paths shown above. Which flight path, A or B, do you think that the crows used most? Why?

   _____

   _____

2. What factors influence the height from which the crows choose to drop the whelks?

   _____

   _____

3. What classroom experiment could model the dropping of whelks to collect and analyze data? What questions would you attempt to answer in your experiment?

   _____

   _____

4. How would the relationship between the number of drops and the height of the drops help you answer your questions?

   _____

   _____

5. Sketch a possible graph of the number of drops required to break a whelk's shell as a function of the height of the drop. What features of your graph can be determined by the context? What assumptions have you made in the graph?

   _____

   _____

One factor in the amount of work done by the crows is the number of times that they need to drop the whelk to break its shell. In the first part of this investigation, we use a simulation to determine the relationship between the height of the drop and the number of drops necessary to break open the shell. For our simulation we will drop peanuts from various heights to collect relevant data.

1. *Conducting the experiment:* To model the dropping of whelks, get a meterstick and a cup of whole, blanched peanuts that have been removed from their shells. Start with a height of 15 cm. Repeatedly drop a peanut until it breaks into two pieces. Record the number of drops needed for the peanut to break. Repeat this experiment for eight peanuts at this same height. Find the mean number of drops required to break open a peanut. Repeat this experiment for heights of 20, 25, 30, 35, 40, 45, 50, and 60 centimeters. You may want to pool your data with data obtained by other groups in the class. Record the data in a table similar to the following.

Height of Drop

Nut	15 cm	20 cm	25 cm	30 cm	35 cm	40 cm	45 cm	50 cm	60 cm
1									
2									
3									
4									
5									
6									
7									
8									
Mean									
Std. dev.									

Number of drops

2. Looking for patterns: What do you see in your data? Do any patterns emerge? How precisely can you express the patterns? Do these patterns fit what you thought would happen? Look for patterns that support or refute the conjecture that you made in question 5, sheet 1, about the graph. _____

_____

_____

_____

_____

1. *Conjecturing a model:* Let H be the height of the drop and N be the number of drops needed.

   a) What does the shape of the graph suggest about a possible equation relating the two variables?

   _____

   b) What is the minimum number of drops required to break open a peanut? How does the number of drops required to break open a peanut vary as the height of the drop increases? What does this information suggest about the form of an equation to model your data?

   _____

   c) Does a minimum height exist at which a peanut will break? Explain your thinking. What does your response suggest about the form of an equation to model your data?

   _____

   d) Use your responses to questions 1(a)–1(c) to explain why a reasonable equation to model your peanut data might be of the form

   $$N = 1 + \frac{k}{H - H_{min}}$$

   _____

2. *Finding and testing the model:* To verify the conjecture, the equation is rewritten to express a linear relationship. The data are transformed in a similar manner to see whether a linear relationship actually exists.

   a) Rewrite the equation in question 1(d) to find $(N - 1)^{-1}$. If the conjectured model is correct, explain why the entries $(N - 1)^{-1}$ should be modeled by a linear expression of the form $mH + b$.

   _____

   b) Using the data from your experiment, complete a table similar to the one shown

   c) Verify that the relationship between $(N - 1)^{-1}$ and H is approximately linear. Find the equation of a line relating $(N - 1)^{-1}$ and H.

Height in cm (H)	Average Number of Drops (N)	$(N - 1)^{-1}$
15		
20		
25		
30		
35		
40		
45		
50		
60		

   _____

   d) Solve your equation in part (c) for the number of drops, N. How well does the equation fit the data?

   _____

   e) Where does this function have a vertical asymptote? According to your data, is this value reasonable for the minimum height needed to break a peanut?

   _____

The work done to break open a peanut depends on the weight of the peanut, the height of the drop, and the number of drops required to break the peanut. The weight of the peanuts is fairly uniform, so the amount of work is essentially a function of the number of drops and the height of the drop.

Work to break a peanut ≈ Number of drops • Height of drop

$$N = 1 + \dfrac{k}{H - H_{min}}$$

Reaching a conclusion: Investigate the amount of work involved to break a peanut.

1. Use your equation for N to create an equation for the work done in breaking open one peanut dropped from a height H.

2. Sketch a graph of your equation.

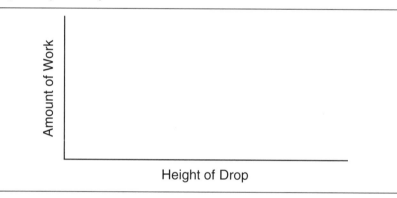

3. Use the graph to decide on a good height from which to drop a peanut. Explain.

   _____

   _____

4. Write a paragraph as you reflect on the mathematics that you have completed. In your paragraph you may want to answer some of the following questions:
   • What kind of equations did you find?
   • What methods did you use to find the equations?
   • What properties of the equations were essential to the process of modeling the peanut data?
   • How does your answer to question 3 relate to your original questions on sheet 1, question 3?

   _____

   _____

   _____

   _____

   _____

   _____

Biologist Reto Zach studied the behavior of the northwestern crows. He conjectured that the crows select a height that minimizes the work needed to break open the whelks. The work done by the crows depends on the weight of the whelk, the height of the drop, and the number of drops required to break open the shell. The weight of the whelks is fairly uniform, since crows feed on only the largest whelks. Thus, the amount of work is essentially a function of the number of drops and the height of the drop. To investigate the work required to break open a shell, Zach dropped a number of whelks of various sizes from a range of heights. The resulting data are given in the graph at the right. The curves are those that Zach sketched by hand.

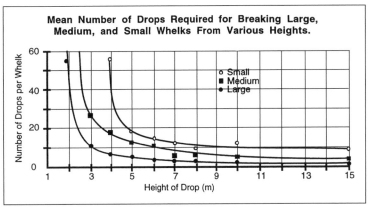

1. After he observed that the crows feed only on the largest whelks, why do you think that Zach also investigated the number of drops needed to break open small- and medium-sized shells?

   _____

   _____

2. What do you observe about the vertical and horizontal asymptotes for the different sizes of whelks? Explain why the asymptotes might differ for the different sizes of whelks.

   _____

   _____

3. Using the data that Zach obtained for large whelks, find an equation for the amount of work required to drop a large whelk from a set height until its shell breaks.

4. Graph your equation for the amount of work required to break open a large whelk. What do you observe about the amount of work as a function of the height?

5. If you were a northwestern crow, from what height would you drop a large whelk to break open the shell? Justify your answer.

   _____

   _____

6. Discuss your results as they relate to Zach's conjecture that dropping whelks from a height of 5 meters is an example of optimal foraging.

   _____

   _____

Discovering an Optimal Property of the Median

Neil C. Schwertman

November 1999

Edited by A. Darien Lauten, Rivier College, Nashua, New Hampshire

TEACHER'S GUIDE

To save time and money, the Postal Service has recently begun using central delivery of mail in new subdivisions. The placement of the central mailboxes in a subdivision or the location of the central depot in any delivery system poses an important and practical problem. To improve their efficiencies, companies desire to select an optimal placement of their distribution facilities.

For example, a gasoline company would like to determine the location of a distribution center for its gasoline trucks. Suppose that n stations must be serviced along a highway and that the trucks can carry only enough gasoline for one station at a time. Since the cost of transporting gasoline from the center to the various stations is quite expensive and is proportional to the distance traveled, the company would like to select a location that minimizes the total distance, that is, the sum of the absolute values of the distances of the stations to the distribution center, that the gasoline must be transported.

The solution to a special case in which all the locations are along a road laid out like a number line can be used to discover an interesting, unique property of the statistical median. This optimal property, which is well known to statisticians but less widely known to mathematicians, has some very practical applications. The following classroom activity demonstrates that for a set of numbers, the sum of the absolute deviations about any point is a minimum when the point chosen is the median of the numbers. This idea gives the solution of the gasoline-company distribution problem.

> *The activity is easy enough for prealgebra students*

The objective of this activity is to encourage students to use reasoning and quantitative problem-solving skills to discover the solution to an important practical problem. By plotting the various absolute-value expressions, the students can practice their plotting skills, as well as see the purpose of the absolute-value function, which some students may view as just another mathematical manipulation. The activity should promote a better understanding that situations occur in which absolute-value functions have real-world meaning. In the gasoline-distribution context, for instance, we look at $|3 - x|$ because we care only how far 3 is from x but not which side of x it is on.

The activity is easy enough for prealgebra students and is also suitable for algebra, geometry, and beginning statistics students; or it can be used to indicate an outline of a proof for more advanced students.

The activity is divided into segments that can be used separately. The introductory activity encourages the students, perhaps working in small groups, to discuss the problem and to articulate their ideas about finding a solution. Each group could use this activity for brainstorming, and the students could share their results and methods, either orally or in writing. This activity should give the students insight into the problem and some direction as they refine their own ideas when they do the main, more formal classroom activities.

The subsequent segments consider two-person, three-person, four-person, and five-person scenarios. Through the progression, the students practice the important problem-solving skill of working from the simplest case to the more complex cases until a pattern emerges and the general case can be deduced.

Some teachers may find the various scenarios a bit repetitious for their students and can omit some of the cases. Others may find the repetition helpful to their students in the discovery process. The general case, in particular, may be too difficult for prealgebra and even algebra classes and could be omitted. If so, the teacher can replace the subscripted x's with letters to simplify the notation and avoid the confusion that beginning students might experience with subscripts. If the general-case activity is used, however, the subscript notation is very useful.

A graphing calculator can be helpful but is not essential in graphing the total absolute deviations in the activity. Throughout the exercise, the students

Neil Schwertman teaches at California State University—Chico, Chico, CA. His interests include developing classroom activities for collaborative learning in introductory statistics and general education classes.

are asked to graph several equations with absolute-value terms. The following equations for the graphing calculator are in the order in which they appear in the exercises:

$Y = \text{abs}\,(3 - x) + \text{abs}\,(20 - x)$
$Y = \text{abs}\,(3 - x) + \text{abs}\,(20 - x) + \text{abs}\,(6 - x)$
$Y = \text{abs}\,(3 - x) + \text{abs}\,(20 - x) + \text{abs}\,(6 - x) + \text{abs}\,(12 - x)$
$Y = \text{abs}\,(3 - x) + \text{abs}\,(20 - x) + \text{abs}\,(6 - x) + \text{abs}\,(12 - x) + \text{abs}\,(8 - x)$

To enrich the classroom activity, the teacher can use a board, nails, fishing line, and a sturdy ring to build a model to demonstrate the result physically, as shown in figure 1. The teacher can ask students to use their intuition to speculate which point or points are the closest to the points represented by the nails. The length of fishing line through the ring represents the total distance—actually, twice the distance—of the ring to each nail. When all possible slack has been removed by pulling the fishing line, the ring is at a location that is closest to the fixed points represented by the nails. Various subsets of the nails can be used. In addition to guessing where the ring will go, one student of a pair can hold the board while the other pulls out all the slack. The model allows students to test their intuition as they observe the optimal point under various conditions and, perhaps, to recognize the pattern and deduce the mathematical solution.

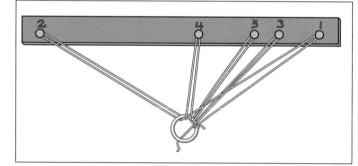

Fig. 1. Diagram of the model

From the graphs, the students should observe the following as p, the location of the chosen point on the number line, increases: in the regions where the slope is negative, the total absolute deviation (TAD) is decreasing; when the slope is zero, the TAD is not changing; and when the slope is positive, the TAD is increasing. This result should reinforce students' understanding of slope as measuring the rate of change. Furthermore, the students should observe that the slopes are negative to the left of the point or region where the TAD is a

minimum and positive to the right of that point or region. They should also see that the minimizing point has equal numbers of points, or x's, on each side and hence is the statistical median.

SOLUTIONS

Sheet 1

1. For the two locations 3 and 20, the students should recognize that the optimal location of the post office—that is, the location that minimizes the sum of the distances to each home—would be any point between 3 and 20 inclusive. The students should also recognize that outside the points from 3 to 20, they could do better by moving the location toward those points. The residents may want the post office to be equidistant from their homes, but that consideration has no bearing on the optimal solution.

2. For the three locations 3, 6, and 20, the students should first recognize from the two-location problem that every point between 3 and 20 minimizes the sum of the distances to those two locations. Since 6 is in that region, it must minimize that sum. But no point is closer to 6 than 6 itself, so the location at 6 minimizes the sum of the distances to the three points.

3. A similar approach can be used for more than three houses. For four houses, any location between the two innermost houses is an optimal location for all four houses. In general, any location that has equal numbers of houses on either side is optimal.

4. Although the distances from town were always integers in the illustrations, these values were chosen for convenience. The same analysis applies to houses located at nonintegral distances from town.

Sheet 2

1. The TAD will vary, depending on the value of p used by the student.

2. The answer depends on the value of p used by the student.

3. Any value of p with $3 \le p \le 20$ has the minimum TAD = 17.

Using Activities from the *Mathematics Teacher* to Support *Principles and Standards*

4. See figure 2.

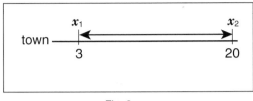

Fig. 2

5. They might meet at any point between the two houses or at either house.

6. See figure 3.

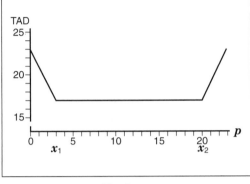

Fig. 3

7. The graph consists of three connected portions of straight lines with different slopes. When $p < 3$, the slope is -2; when $3 < p < 20$, the slope is 0; and when $p > 20$, the slope is 2. The slope increases by 2 whenever p increases and crosses one of the x's.

8. Whenever $p < 3$, moving p to the right a distance d moves p closer to each of the two houses by d, thus decreasing the TAD by $2d$. When $3 < p < 20$, moving p to the right does not change the TAD, because p moves closer to the house at 20 but further away from the house at 3 by the same amount. When $p > 20$, moving p to the right a distance d moves p away from each of the two houses by a distance d, thus increasing the TAD by $2d$.

Sheet 3

1. See figure 4.

2. The graph consists of four connected portions of straight lines. The portions to the left of 6 have a negative slope, whereas those to the right have a positive slope. For $p < 3$, the slope is -3; for $3 < p < 6$, the slope is -1; for $6 < p < 20$, the slope is 1; and for $p > 20$, the slope is 3.

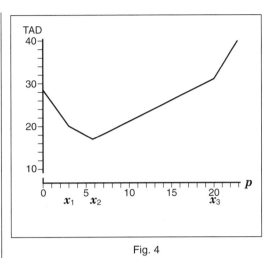

Fig. 4

3. Each time that p increases and crosses one of the locations, or x's, the slope increases by 2.

4. Slopes measure the change in total absolute deviation (TAD) as the point p is moved one unit to the right. When p is to the left of all three points, moving p one unit to the right moves p one unit closer to all three points, or houses; so the distance decreases by 3 and hence the slope, which measures the change, is -3. When p is between the houses at x_1 and x_2, moving p one unit to the right moves p one unit closer to the two houses at the largest distances from town but one unit farther away from the house closest to town. Thus, the total distance decreases by 1 and the slope is -1. Similarly, when p is between the houses at x_2 and x_3, moving p one unit to the right moves p one unit closer to the house farthest from town and one unit farther from the houses located at x_2 and x_3. The slope is then 1.

Notice that the slope is negative, that is, the TAD is decreasing, until the same number of houses are on either side of p. At that point, moving p in either direction moves p closer to half the houses but farther from the other half by the same amount, and the change in the TAD is 0. When more houses are on one side of p than on the other, the TAD can be decreased by moving p toward the side with the greater number of houses.

5. The minimum TAD occurs at 6.

6. Yes; it is the only point that minimizes the TAD.

7. To the left of 6, the slope in figure 4 is negative, indicating that the TAD is decreasing, whereas to the right the slope is positive, showing that the TAD is increasing. Thus, the TAD is decreasing until $p = 6$.

8. See figure 5.

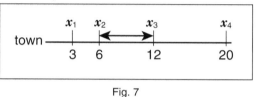

Fig. 5

Sheet 4

1. See figure 6.

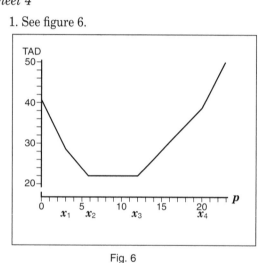

Fig. 6

2. The graph consists of five connected portions of straight lines with the slope changing at the location of each house. The slope is decreasing when $p < 6$, unchanged when p is between 6 and 12 inclusive, and increasing when p is greater than or equal to 12. The slope is -4 when $p < 3$, -2 when $3 < p < 6$, 0 when $6 < p < 12$, 2 when $12 < p < 20$, and 4 when $p > 20$.

3. As before, slopes measure the change in total absolute deviation (TAD) as the point p is moved one unit to the right. Observe that when p is to the left of all four points, moving p one unit to the right moves p one unit closer to all four points, or houses; that is, the distance decreases by 4, and hence the slope, which measures the change, is -4. When p is between the houses at x_1 and x_2, moving p one unit to the right moves p one unit closer to the three houses at the largest distances from town but one unit

farther away from the house closest to town. Thus, the total distance decreases by 2 and the slope is -2. Similarly, when p is between the houses at x_2 and x_3, moving p one unit to the right moves p one unit closer to the two houses farthest from town and one unit farther away from the houses located at x_1 and x_2. The slope is then $-2 + 2$, or 0. Notice that the slope is negative, that is, the TAD is decreasing, until the same number of houses are on either side of p. At that point, moving p in either direction moves p closer to half the houses but farther away from the other half by the same amount, and the change in the TAD is 0. When more houses are on one side of p than on the other, the TAD can be decreased by moving p toward the side with the greater number of houses.

4. See figure 7.

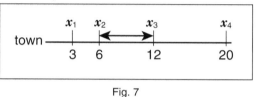

Fig. 7

Sheet 5

1. See figure 8.

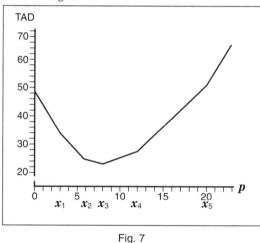

Fig. 7

2. The graph consists of six connected portions of straight lines with differing slope. The slope is decreasing to the left of 8 and increasing to the right of 8. The slope is -5 when $p < 3$, -3 for $3 < p < 6$, -1 for $6 < p < 8$, 1 for $8 < p < 12$, 3 for $12 < p < 20$, and 5 for $p > 20$. The slope increases by 2 when p increases and crosses an x.

3. Again, slopes measure the change in the TAD as the point p is moved one unit to the right. Observe that when p is to the left of all n points, moving p one unit to the right moves p one unit closer to all n points, or houses; that is, the distance decreases by n, and hence the slope, which measures the change, is $-n$. When p is between the houses at x_1 and x_2, moving p one unit to the right moves p one unit closer to the $n - 1$ houses at the largest distances from town but one unit farther away from the house closest to town. Thus, the total distance decreases by $(n - 1) - 1$, and the slope is $-n + 2$, which is -3 if n is 5. Similarly, when p is between the houses at x_2 and x_3, moving p one unit to the right moves p one unit closer to the $n - 2$ houses farthest from town and one unit farther away from the houses located at x_1 and x_2. The slope is then $-(n - 2) + 2 = -n + 4$, which is -1 if n is 5.

Notice that the slope is negative, that is, the TAD is decreasing, until the same number of houses are on either side of p. At that point, moving p in either direction moves p closer to half the houses but farther away from the other half by the same amount, and the change in the TAD is 0. When more houses are on one side of p than on the other, the TAD can be decreased by moving p toward the side with the greater number of houses.

4. The minimum TAD occurs at $p = 8$. At that point, equal numbers of houses are on each side. Moving p from that point moves p closer to less than half the houses and farther away from more than half the houses, thus increasing the total TAD.

5.

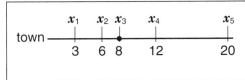

6. The point is called the median.

TABLE 1	
n	Point that Minimzes the TAD
2	$\dfrac{x_1 + x_2}{2}$
3	x_2
4	$\dfrac{x_2 + x_3}{2}$
5	x_3
6	$\dfrac{x_3 + x_4}{2}$
7	x_4
8	$\dfrac{x_4 + x_5}{2}$
9	x_5
n(odd)	$x_{(n+1)2}$
n(even)	$\dfrac{x_{(n/2)} + x_{(n/2)+1}}{2}$

CONCLUSION

Doing these activities outlines a proof that the point that minimizes the total distance to a set of locations on a straight road has an equal number of locations on either side. The median, the midpoint of any interval solution, is the solution to the meeting problem, as well as the solution to such problems as the location of the central depot for a distribution system.

In a small community called Lineville, the homes are along a straight road, which is laid out like a number line. The residents must select the location along the road for a small post office where they will pick up their mail. To make the location as convenient as possible, the residents decide to place the post office so that the sum of the distances between it and the homes is as small as possible. First, suppose that only two homes were on the road, one located at 3 on the number line and the other located at 20, as shown here.

1. What would be the best location? How did you determine that this location is the best one?

Next, suppose that three houses are on the road, one each at points 3, 6, and 20.

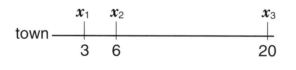

2. Where would the best location for the post office be?

3. Experiment with a few locations, and determine the best place for a post office. What did you do to find it? How did you determine that it is the best location? Did tables and graphs help? Be prepared to share your results and methods with the rest of the class. Could you use your methods to find the best location for the post office if the number of homes is four, five, or some other number?

4. In these illustrations, we used an integer for distances from town. Would the same problem-solving approach apply if the distance from town were not integers like 3, 6, and 20 but numbers like 3.1, 6.2, and 19.7?

A group of people living along a straight road wants to choose a place along the road to meet. They want to select the place on the road that minimizes the total distance that they must travel, perhaps to minimize gasoline consumption. The road runs in an east-west direction, and all the people live to the east of some town. To create a mathematical model of the problem, think of the road as a number line, with the town located at 0. Each person's home is represented by a positive numerical coordinate that indicates how far east of the town that person lives.

If the number of people is n, arranging these coordinates in ascending numerical order, $x_1 \leq x_2 \leq \ldots \leq x_n$, is helpful. To develop the idea, first use the simplest case, that in which n is 2; that is, the two homes are located at x_1 and x_2. Let p be any point. The point p should be chosen so that the sum of the distances—that is, the distance between x_1 and p plus the distance between x_2 and p—is as small as possible. Since all distances are considered to be positive, each distance can be represented as an absolute value, so the sum of the absolute values of the differences between the x's and p should be a minimum.

For example, suppose that the first person lives 3 km from town and that the second person lives 20 km from town, and that they decide to meet at a point along the straight road 22 km from town. Their distances from town are represented by $x_1 = 3$, $x_2 = 20$. The meeting point is $p = 22$, as shown at the right.

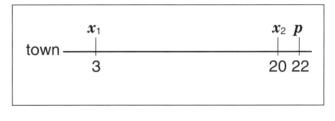

The total absolute deviation (TAD) is the sum of absolute values. For this example, the TAD is $|3 - 22| + |20 - 22| = 19 + 2 = 21$.

Choose a value for p other than 22, and find the TAD for that p. That is, find the value of $|3 - p| + |20 - p|$ for your p.

1. What is your TAD? _____

2. How does your TAD compare with the TAD when $p = 22$? _____

3. Try other values for p until you find one that you are sure will make the TAD as small as possible. What value did you obtain for p? _____ What minimum value did you obtain for the TAD? _____

4. Indicate on the previous number line all choices for p that minimize the TAD.

5. What does question 4 tell you about where the two people might meet to minimize their combined travel distance?

Creating a graph that shows how TAD changes as you change p can help you visualize what is happening.

6. Either by hand or by using a graphing calculator, graph TAD versus p. The chart shown at the right of the graph may be helpful, particularly if you are graphing by hand. Fill in as many values for p as you need to complete the graph. Think about the values of p that will supply the most information about the graph. Include at least a point or two in each of the three regions: $x < x_1$ where $x_1 = 3$; $3 \le x \le 20$; and $x > x_2$ where $x_2 = 20$.

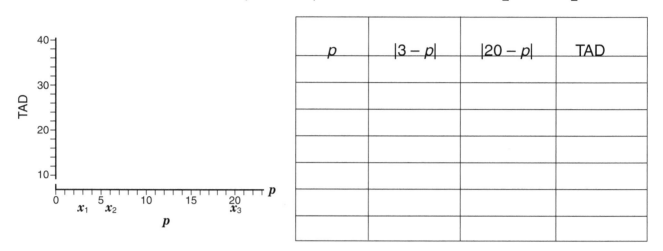

p	$|3 - p|$	$|20 - p|$	TAD

7. How is the graph different in each region?

What is the slope when p is less than 3? _____ Between 3 and 20? _____

Greater than 20? _____ How much does the slope of each segment change as p increases and crosses one of the x's?

8. How can you use the travel-distance scenario to explain the change in slope from region to region?

Something to consider:

How does the TAD change as p increases? Why?

Suppose that a third person, who lives 6 km from town on the same straight road, joins the first two. The number of locations, or x's, is 3 ($n = 3$), as shown. Again, for each choice of p, the TAD is a sum of absolute values. For instance, if $p = 10$, the TAD is $|3 - 10| + |6 - 10| + |20 - 10| = 21$.

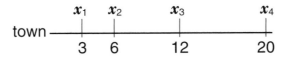

Creating a graph that shows how the TAD changes as you change p can help you visualize what is happening.

1. Either by hand or by using a graphing calculator, graph TAD versus p. Again, the chart at the right of the graph may be helpful. Fill in as many values for p as you need to complete the graph. Think about where the slope of the graph might change, and include values for p to verify where it changes.

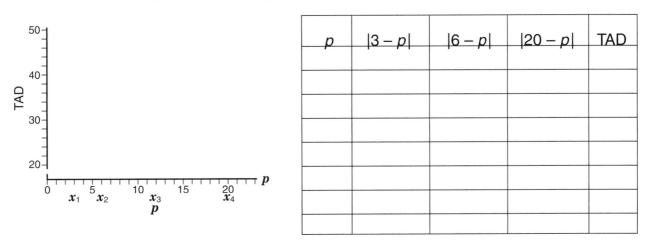

p	$|3 - p|$	$|6 - p|$	$|20 - p|$	TAD

2. How would you describe the graph? _____
   What is the slope when p is less than 3? _____ Between 3 and 6? _____
   Between 6 and 20? _____ Greater than 20? _____

3. How much does the slope change as p increases and crosses one of the x's?

4. How can you use the travel-distance scenario to explain the change in slope from region to region?

5. Where does the minimum TAD occur?

6. Is that point the only point that minimizes the TAD?

7. How do you know?

8. Indicate on the line where the minimum occurred.

Next, suppose that a fourth person joins the group. That person lives 12 km from town on the same straight road. The number of x's is 4 ($n = 4$), as illustrated on the line.

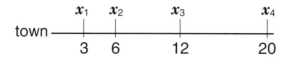

1. Either by hand or by using a graphing calculator, graph TAD versus p. Again, you may find the chart helpful.

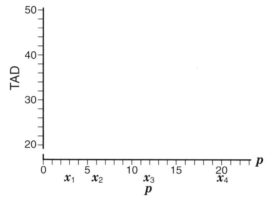

p	$|3 - p|$	$|6 - p|$	$|12 - p|$	$|20 - p|$	TAD

2. Describe the graph.

   What is the slope of the straight-line segments in the graph when p is less than 3? _____ Between 3 and 6? _____ Between 6 and 12? _____ Between 12 and 20? _____ Greater than 20? _____

3. Use the travel-distance scenario to explain these changes of slope.

4. On the line, show where the TAD is a minimum.

From the *Mathematics Teacher*, November 1999

Finally, suppose that a fifth person, who lives 8 km from town on the same road, joins the group. The number of x's is now 5 ($n = 5$), and their distances from town are shown on the following line.

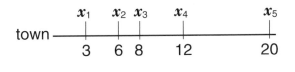

1. Either by hand or by using a graphing calculator, graph TAD versus p. You might use a chart like those used in previous examples.

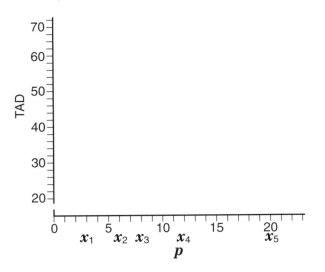

2. Describe the graph.

   What is the slope when p is less than 3? _____
   Between 3 and 6? _____ Between 6 and 8? _____ Between 8 and 12? _____ Between 12 and 20? _____ Greater than 20? _____ How much does the slope change as p increases and crosses one of the x's? _____

3. Use the travel-distance scenario to explain the changes in slope.

4. Where does the minimum TAD occur? _____ Explain why.

5. On the line, indicate where the minimum TAD occurs.

   Observe that when the number of x's that represent the distances from town is odd, only one point minimizes the TAD. When the number of x's is even, an interval usually exists where many values of p minimize the TAD.

   If the midpoint of that interval is used, then the number of locations, or x's, greater than and less than the point p, are the same in either situation.

6. What is such a point called in statistics?

From the previous activities, a pattern has begun to emerge. Suppose that n persons live along a straight road with distances from town of $x_1 \leq x_2 \leq x_3 \leq \ldots \leq x_n$. When n is odd, a single point minimizes the TAD, but when n is even, an interval usually exists. Rather than specify an interval, use the midpoint of the innermost interval when n is even for convenience. On the basis of this convention, the points that minimize the TAD for $n = 2, 3, 4,$ and 5 are given in the following chart.

n	Point That Minimizes TAD
2	$\dfrac{x_1 + x_2}{2}$
3	x_2
4	$\dfrac{x_2 + x_3}{2}$
5	x_3
6	
7	
8	
9	
n (odd)	
n (even)	

Think about the process that you used to find the point or points that minimize the TAD and the pattern that it developed. What caused this point to minimize the TAD?

Use your reasoning to deduce the point that minimizes the TAD when the number of persons is increased for any number of persons n. Fill in the rest of the table to see the pattern that has been developed.

By doing these activities, you have outlined a proof that the point that minimizes the total distance to a set of locations along a straight road has an equal number of locations on either side. If you take the midpoint of any interval solution, that point is called the median. That point is the solution to the meeting problem, as well as the solution to such problems as the location of the central depot for a distribution system.

Will the Best Candidate Win?

Teresa D. Magnus

January 2000

Edited by Darien Lauten, *Rivier College, Nashua, New Hampshire*

> *The plurality method can produce a winner who is liked least*

Students think that they are familiar with the concept of voting. After all, they have heard about governmental elections, Academy Award voting, and the ranking of football teams. They have probably participated in club and school elections. Yet if you ask students about voting methods, most can describe only one voting method, namely, plurality. Not only have they never questioned its fairness, but considering other methods is new for them.

The following activities allow students to explore alternative voting methods. They discover what advantages and disadvantages each method offers and also see that each fails, in some way, to satisfy some desirable properties. They are particularly surprised to discover that the plurality method can produce a winner who is liked least by a majority of the voters. In addition, students look at how elections can be manipulated. One extension involves discussing Arrow's impossibility theorem, which states that it is impossible for a voting system to satisfy all the desirable features. Although these activities, including the statement and proof of Arrow's theorem, require only basic arithmetic, they allow students to engage in high-level mathematical thinking.

This activity lends itself easily to interdisciplinary instruction. Current events on the national, local, or school level can be incorporated into the project. If students are involved in making such group decisions as choosing a class gift or service project, selecting a time to hold an event, or arranging for refreshments, relevant alternatives can be substituted for the activities on the worksheets.

Grade level: 8–14

Objectives: This activity helps students see connections between mathematics and such other disciplines as government, history, ethics, and sports. Students develop skills in mathematical reasoning and apply those skills to everyday situations. They learn about various voting methods, ways in which these methods can be manipulated to achieve certain outcomes, and the impossibility of a fair election when more than two alternatives are available.

Directions

To introduce the topic and to familiarize students with the table format, I recommend the following large-group activity. Have the students suggest activities or destinations for a hypothetical class trip, and write the first three suggestions on the chalkboard. Ask each student to list her or his first, second, and third choices on a piece of paper, permitting no tied rankings.

Suppose that your students suggested archery (A), biking (B), and canoeing (C). You would then create a table, similar to table 1, where the columns represent all the possible preference lists. At the top of each column, write the number of students who ranked the options in the order given. Ask students which alternative wins and how they determined the winner. If students generate only one method of tallying the winner, ask them to think of a different way.

Each activity sheet is designed for groups of three to four students.

TABLE 1					
Example Preference Table					
First choice	A	A	B	B	C	C
Second choice	B	C	A	C	A	B

Sheet 1

Plurality voting is the method most familiar to students. In this method, each member is given one vote and the option that receives the most votes wins. Variations of the plurality method are used in choosing state representatives and senators, ratifying proposals, and selecting Academy Award winners. When a candidate receives more than 50 percent

---

Teresa Magnus teaches at Rivier College, Nashua, New Hampshire. Her interests include abstract algebra, geometry, and mathematical history. Cooperative learning and writing assignments play an important role in her classes.*

of the vote, including situations in which only two candidates are being considered, then plurality does produce a preferred candidate. However, in many situations, plurality may not produce a clear preference.

Since students are comfortable with the plurality method, they can usually complete this activity sheet quickly in small groups. They are often surprised to find that skiing comes in both first and last and often mention their surprise in their responses to question 3. A reasonable answer for question 4 would be the speed and ease of the plurality method. If students get stuck on question 5, ask them to think about how winners are determined in sports tournaments or when many alternatives are available.

A nice follow-up to this activity sheet is a discussion about variations of the plurality method, including runoff elections, the electoral college, and two-thirds majority. Another variation of plurality, the Hare system of voting, involves a series of elections. At each stage the option, or options, with the least number of votes are eliminated from future ballots. The voters who originally voted for the eliminated option vote for the remaining option that they have ranked highest. I find that this point is worth emphasizing because students often tend to overlook these voters.

The second part of this activity sheet introduces students to three voting methods: the Hare system, Borda count, and sequential pairwise voting. Although the sheet gives instructions for each voting method, some student groups may need help following the directions. With the Borda count, I show students how they can put a 0 next to third place, a 1 next to second place, and a 2 next to first place on the chart as an aid to totaling the points.

A quick way for students to verify that their totals are reasonable is to determine the total number of votes, that is, three points per voter times forty voters, and verify that this result matches their total number of points.

Comparing the technique of sequential pairwise voting with a single-elimination tournament, with byes, may help students understand the method, but care should be taken to distinguish between the two. In sequential pairwise voting, the pairings are sequential and no simultaneous pairings take place. Sketching the corresponding "tournament bracket" diagrams as a demonstration may help clarify the distinction.

Ties do not occur in the problems given; however, the instructor should know how they are handled in the multistep voting systems. With the

The Hare system of voting involves a series of elections.

Hare system, whenever two or more options share the least number of votes, they are eliminated at the same time. If all remaining candidates have the same number of votes, none are eliminated; they are all considered tied for the win. Whenever two candidates tie during a head-to-head contest in sequential pairwise voting, neither is eliminated; they both continue and compete in a three-way contest with the next candidate.

Have students discuss the advantages and disadvantages of each of these voting methods, first in their small groups and then with the entire class.

Answers to sheet 1

6. In the first round of voting, skiing gets seventeen votes, rafting gets eleven, and caving gets twelve. Thus, rafting is eliminated. In the follow-up election, skiing gains one of rafting's votes, for a total of eighteen, whereas caving gets its original twelve votes plus ten votes from rafting, for a total of twenty-two. Skiing is now eliminated, leaving caving as the winner.

7. Caving gets 2(12) + 17, or 41, points. Rafting gets 2(11) + 18, or 40, points. Skiing gets 2(17) + 5, or 39, points. Caving wins using the Borda count.

8. In the first vote, skiing gets eighteen votes to caving's twenty-two. Thus, skiing is eliminated and caving meets rafting in a head-to-head contest. This time, caving gets nineteen votes, whereas rafting gets twenty-one. Rafting wins.

9. Answers may vary; however, many students will assert that rafting gets an unfair advantage in problem 3.

10. A hint may be needed here. One possible answer is to eliminate skiing, the activity that the greatest number of voters ranked last, and then to hold a runoff election between the remaining options.

Sheet 2

Here students investigate how a voter or block of voters can influence the results of an election by submitting a ballot that does not represent their true preferences. Although the terminology is avoided on these sheets, each of the problems on sheet 2 demonstrates that a property known as *independence of irrelevant alternatives* does not hold for these voting methods. In other words, a losing candidate can win the election without any voters' having moved the new winner ahead of the original winner in their preference lists. They may

Using Activities from the *Mathematics Teacher* to Support *Principles and Standards*

have moved other, irrelevant, candidates above or below one of these two. As an example, consider the following chart of preference lists:

		Number	
Ranking	2	3	4
First choice	A	B	C
Second choice	B	A	A
Third choice	C	C	B

Candidate C wins using the plurality method. However, if the two people represented by the first column switched the positions of A and B in their preference list, B would win by plurality. Note that the order of B and C was not reversed. By moving B ahead of an irrelevant alternative, A, on two preference lists, B was able to win.

Question 1 on sheet 2 is adapted from *Introductory Graph Theory* (Chartrand 1985, 168).

Answers to sheet 2

1. An editor who was voting according to his or her true preferences would probably rank his or her school first and Big City High second, or vice versa. In this problem though, students should discover that another strategy benefits the editor's school. By ranking his or her school's team first and not including Big City High among the top ten, the editor's school gets 9(9) + 10, or 91, points compared with Big City's 9(10) + 0, or 90, points. Give some credit to groups who create a tie, but point out to them that they need not include Big City High in the top ten.

2. a) If the plurality method is used, A wins, with 48 percent of the vote compared with B's 28 percent and C's 24 percent.
   b) If the voters in this group ranked B ahead of C, then B would win instead of A.
   c) If the Hare system of voting is used, then C is eliminated first. In the next round, B wins with 52 percent of the vote. The last 10 percent of the voters would be most disappointed in this result. If they submitted a ballot with the ranking C, A, B, then B would be eliminated in the first round and A would beat C.

Sheet 3

At first, one would expect that if a candidate, called a Condorcet winner, could beat each of the other candidates in head-to-head contests, that candidate should win the election in which all candidates compete. Students are surprised to discov-

er that this so-called Condorcet-winner criterion does not hold for the plurality method, Borda count, and Hare system. Students should be asked to explain why it does hold for sequential pairwise voting.

In sheet 3, tournament digraphs are used to help students visualize the results of pairwise voting. In question 1, most students expect candidate B to win in every method. They discover in problem 2, which has as its solution the digraph in 1, that although B may turn out to be the winner under each of these voting methods, they can be positive that B wins only in sequential pairwise voting.

Problem 3 is tangential to the main topic. You may wish to omit it or use it as a take-home bonus question. With n candidates,

$$\frac{n(n-1)}{2}$$

A Condorcet winner might lose an election under the Hare system of voting

arrows are involved. Some students will write this expression in the form $1 + 2 + 3 + \ldots + (n-1)$. One way of deriving the first formula is by noting that n points exist and that each point has an arrow to or from each of the other $n - 1$ points. Since each arrow touches two points, the number of arrows is

$$\frac{n(n-1)}{2}.$$

If students used patterns to discover the formula $1 + 2 + 3 + \ldots + (n-1)$, you can show them that

$$1 + 2 + 3 + \ldots + (n-1)$$
$$= \frac{1 + 2 + 3 + \cdots + (n-1)}{2}$$
$$+ \frac{((n-1) + \cdots + 3 + 2 + 1)}{2}$$
$$+ \frac{1}{2}\left(1 + (n-1)\right) + \left(2 + (n-2)\right) + \left(3 + (n-3)\right)$$
$$+ \cdots + \left((n-1) + 1\right) = \frac{1}{2}(n-1)n.$$

Sheet 3 also introduces the term *Condorcet winner*. Problem 6 challenges students to be more creative and develop their own examples of tables of preference lists that show that the Hare system does not satisfy the Condorcet-winner criterion. Make sure that students understand that their tables of preference lists for this problem must produce a Condorcet winner. Suggest that students think about how a Condorcet winner might lose an election under the Hare system of voting. Note that the Condorcet winner must be eliminated at an early stage. It cannot have a lot of first-place votes. Students can later experiment with different tables of preference lists to create this situation.

Answers to sheet 3

4. The tournament digraph follows:

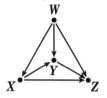

The exact arrangement of the vertices W, X, Y, and Z is not important. To check students' graphs, verify that exactly one arrow appears between every pair of vertices, three arrows leave W, three arrows point toward Z, and an arrow goes from X to Y.

5. By using the Borda count, W gets $3(3) + 2(2) + 2 = 15$ points, X gets $3(4) + 2(3) = 18$ points, Y gets $2(2) + 7 = 11$ points, and Z gets $3(2) + 2(2) = 10$ points, so X wins. Have students verify that they have calculated the right number of total points for each option.

6. One example is given by the following table:

2	3	3
W	X	Y
X	W	W
Y	Y	X

Here, W is a Condorcet winner that gets eliminated in the first round of Hare voting.

7. Since sequential pairwise voting involves only head-to-head contests, a Condorcet winner will win every contest it is in and hence wins the election.

What is next?

Now that students have discovered that each of these voting methods may not produce the expected result, a new question arises. Does a voting system exist that satisfies all desirable criteria? Kenneth Arrow proved that not only do these four voting methods not satisfy both the Condorcet-winner criterion (CWC) and independence of irrelevant alternatives (IIA) but that it is impossible to create any voting system that does. To be more precise, any voting system that always produces at least one winner cannot satisfy both CWC and IIA. The proof of this theorem, which can be found in Brams et al. (1996, 426–30), requires an understanding of the topics on these worksheets along with high-level mathematical reasoning. Students have difficulty understanding that if a criterion does not hold for any one table of preference lists under a particular voting method, then the voting method fails to satisfy the criterion. Other preference lists may exist for which there appears to be no conflict.

The upcoming presidential primaries offer another opportunity for extending this activity and linking it with statistics and social studies. Have students survey a sample of adults, asking them to rank the leading Republican—or Democratic, if more than three candidates run—candidates in order of preference. Ask students to create a table of preference lists illustrating their data and to determine the winner on the basis of their data, using each voting system studied. Students can report their sampling methods, calculations, findings, and any discrepancies in an essay or news story.

REFERENCES

Brams, Steven J., et al. "Social Choice: The Impossible Dream." In *For All Practical Purposes*, 4th ed., edited by COMAP, the Consortium for Mathematics and Its Applications, 411–42. New York: W. H. Freeman & Co., 1997.

Chartrand, Gary. *Introductory Graph Theory*. New York: Dover Publications, 1985.

THE PLURALITY METHOD AND OTHER VOTING SYSTEMS Sheet 1A

The forty members of your school adventure club are trying to decide what type of trip to take. The chart shows how the club members rank the three options.

	Number of Voters					
Ranking	10	7	1	10	4	8
First choice	skiing	skiing	rafting	rafting	caving	caving
Second choice	rafting	caving	skiing	caving	skiing	rafting
Third choice	caving	rafting	caving	skiing	rafting	skiing

1. A common method of voting is plurality. In this system, each person casts one vote for a first choice and the option with the most votes wins. On the basis of the chart, which activity would be the winner under the plurality system? _____

2. Which activity is liked least by the largest number of members; that is, which activity is ranked third by the largest number of voters? _____

3. Why might the plurality method not produce results satisfactory to all voters?

4. Why do you think the plurality method is used so often?

5. Think of some variations of plurality voting or other voting techniques that might prove more satisfactory to the voters. Within your group, describe or develop at least two other vote-tallying methods that have not yet been discussed in class.

Use the set of preference lists (chart) from the first part of this activity sheet to answer the following:

6. The *Hare voting system* involves taking an initial poll in which each person casts one vote for his or her favorite option. The option receiving the least number of first-place votes is eliminated, and another poll is taken. Those members who originally voted for the eliminated option vote for their second choice in the runoff election. Continue eliminating the options that receive the fewest votes and repolling until a single winner or tied winners remain. Which activity would the adventure club choose using the Hare system? _____
Describe the process as options are eliminated.

7. Another voting method, called a Borda count, takes into account each voter's first, second, and third choices. Each first-choice vote is awarded two points, each second-choice vote is worth one point, and third choices receive no points. For example, skiing has seventeen first-choice votes and five second-choice votes, for a total of 2(17) + 1(5), or 39, points. Determine the total number of points for the other two activities. Which activity has the most points using this method?

8. *Sequential pairwise* voting involves a sequence of head-to-head contests. The organization first votes on two of the options, and the preferred option is then matched with the next option while the "loser" is eliminated. A club member suggests that the club should first vote between skiing and caving and then have a vote between the winner of that contest and the remaining option, rafting. Which activity is chosen by this method?

9. Which of the methods—plurality, Hare, Borda count, or sequential pairwise voting—is the fairest in this situation? Why? Which is the least fair? Why?

10. Suppose that your preference is rafting. Devise a voting system that would enable rafting to be chosen and that would sound fair to the other club members.

1. Suppose that you are one of the ten sports editors whose votes together determine the rankings of state high school football teams. A variation of the Borda count is used in which each voter ranks his or her top ten teams out of the many teams in the division. Points are awarded as follows:

10 points for each first-place vote
9 points for each second-place vote
⋮
1 point for each tenth-place vote

You suspect that the other nine sports editors will rank Big City High first and your high school's team second. Can you submit a ranking that will enable your team to earn the highest number of points? Explain.

2. Candidates A, B, and C are running for office. The preferences of the voting community are given in the following chart:

	Percent			
Ranking	38%	29%	24%	10%
First choice	A	B	C	A
Second choice	B	A	B	C
Third choice	C	C	A	B

a) Who will win the election if the plurality method is used? _____

b) By still using the plurality method, does a strategy exist that the voters in the 24 percent column could use to get a result more to their liking?

c) Who will win the election if the Hare system is used? _____
Which column of voters would be most disappointed in the result? _____
Devise an alternative ranking that they could submit to get a result more to their liking when the Hare system is used.

We can use a diagram called a *tournament digraph* to illustrate the expected results of head-to-head polls among the candidates. We use vertices (dots) to represent each of the candidates and draw an arrow from one candidate to another if the first candidate would beat the second in a head-to-head competition. If no ties occur, exactly one arrow will join every pair of candidates.

Example: The tournament digraph corresponding to the situation on the first sheet is sketched at the right. Note that in head-to-head matches, skiing beats rafting, rafting beats caving, and caving beats skiing.

1. In the example above, no activity defeated all the other activities in head-to-head voting. Suppose instead that the tournament digraph for a set of preference lists looked liked this:

    Who would you expect to win using plurality? _____

    Hare system? _____

    Borda count? _____

    Sequential pairwise voting? _____

2. Sketch the tournament digraph for the election in problem 2, sheet 2. Does this result agree with your claim in question 1?

3. How many arrows are in a tournament digraph with five candidates? _____ With ten candidates? _____ With *n* candidates? _____ Can you develop a formula? _____

4. Sketch the tournament digraph for the following:

Ranking	Number			
	3	2	2	2
First choice	W	Z	X	X
Second choice	X	Y	Z	W
Third choice	Y	W	Y	Y
Fourth choice	Z	X	W	Z

5. Determine the winner of the election in problem 4 when the Borda count method is used. _____ Is it what you expected? _____ Note that with four candidates, each first-place vote will be worth 3 points, each second-place vote will be worth 2 points, and each third-place vote will be worth 1 point.

6. When a candidate beats every other candidate in head-to-head contests, we call the candidate a Condorcet winner. Develop a table of preference lists for which the Condorcet winner would lose the election under the Hare system of voting.

7. Is it possible to have a Condorcet winner who does not win using the sequential pairwise voting method? _____ Support your answer.

The Jurassic Classroom

Art Johnson

February 2000

Edited by A. Darien Lauten, *Rivier College, Nashua, New Hampshire*

Proportional reasoning is a crucial part of any comprehensive mathematics program. Aspects of proportional reasoning can be introduced to students in primary grades, but most students are not developmentally ready to reason proportionally until upper middle school (Lawton 1993; Hart 1988; Piaget, Inhelder, and Szeminska 1960). A full understanding of proportionality opens the doors to many interesting applications of mathematics. A student who understands the proportionality involved in the areas and volumes of similar solids, for example, can understand why elephants and hippopotamuses have such short, massive legs; why a mouse can fall from a height of several stories and walk away, whereas a human will usually suffer fatal injuries; why insects cannot survive in extremely cold climates; why a bird or small mammal eats nearly its weight in a single day; and why a giant like Paul Bunyan could not exist. As early as 1638, Galileo wrestled with similar questions. In *Dialogues Concerning Two New Sciences*, Galileo wrote,

> . . . if one wants to maintain in a great giant the same proportion of limb as that found in an ordinary man he must either find harder and stronger material for making bones, or he must admit a diminution of strength in comparison with a man of medium stature; for if his height be increased inordinately he will fall and be crushed under his own weight.

These ideas can also help students make a connection between mathematics and natural science.

A major component of proportional reasoning is the study of similarity. Students generally begin to study similar figures in upper elementary school and culminate their study of similar figures in high school geometry. Some studies of similarity include the concepts of expansions and dilations as a means of internalizing the relationships of similar figures and linking similarity to proportional reasoning. If students are to fully understand proportional relationships, especially those involving similar solids, they must construct their own understanding rather than learn to manipulate proportions. For many students, proportional reasoning exists simply and exclusively in algebraic proportion problems (Kaput and West 1994; Cramer, Post, and Currier 1993; Hart 1988), yet proportional reasoning is more than merely using an algorithm to solve for a missing term in a typical problem. In fact, even the ability to solve a proportion by rote cross multiplication does not indicate an ability to reason proportionally (Post, Behr, and Lesh 1988; Lesh, Post, and Behr 1988; Behr et al. 1983). This activity helps students build an understanding of the proportional relationships between similar solids.

Proportional relationships between solid figures are particularly difficult for students in that three ratios must be considered: (1) the ratio between corresponding lengths, (2) the ratio between corresponding areas, and (3) the ratio between corresponding volumes. Students intuitively set all three ratios equal to one another. But they are not equal, a fact that creates a difficulty for students. The ratios are related, but the relationships are exponential, not linear, another source of difficulty. This activity helps students make sense of the three ratios and begin to construct proportional relationships among them.

Classroom Activity

The growable dinosaurs used for this classroom exploration expand when placed in water. The dinosaurs will grow to about six times their original length in twenty-four hours. This activity allows students to discover what this sixfold increase in length means in relation to the dinosaur's increase in area and volume.

When the activity is completed, the dinosaurs can be removed from water, placed in a dish, and allowed to dry out. After about two weeks, the

Art Johnson teaches at Nashua Senior High School, Nashua, New Hampshire. His interests are in mathematics history and geometry at the middle and high school levels. His research investigates how students apply proportional reasoning to geometric situations.

dinosaurs will have shrunk to their original size and can be reused. They come in a variety of types and colors.

Materials: One growable dinosaur for each group, graph paper, centimeter cubes, ruler or tape measure, graduated cylinder, sealable plastic storage bags (about one-gallon size), and water troughs

Student level: This activity is appropriate for students in a typical high school geometry class. With some adjustments, it can also be used with middle school students. The activity requires parts of three days for completion.

Day 1

On the first day, students need to take linear measurements of their dinosaur for sheet 1; an area measurement for sheet 2, question 1; and a volume measurement for sheet 3, question 1. For the linear data, students should select a length that is relatively easy to measure and that represents a reasonable length of the dinosaur, such as the length from head to tail or from wingtip to wingtip. The area of a dinosaur is best determined by having students trace the outline of their dinosaur on the grid on sheet 2. The outline is likely the same shape as a cross section of the dinosaur along its entire length. The growth of the area in this outline will show the same proportional change as the growth in the total surface area.

To determine the dinosaur's volume, students need a graduated cylinder, which can often be obtained from the science department. A 250 cc graduated cylinder is large enough to accommodate any of the "dry" dinosaurs; some dinosaurs fit into smaller graduated cylinders. Students fill the graduated cylinder with water to a specific reading, such as 10 cc, and then record the water level after they place their dinosaur into the water. Since the dinosaurs sink below the water's surface, the volume of the dinosaur is equivalent to the amount of water that the dinosaur displaces. After they measure the dinosaur's volume, the students remove the dinosaur from the water. The concept of water displacement can also be used to measure the volume of the grown dinosaurs.

A second task for this first day is for students to determine a schedule to measure the length of their dinosaur as it grows for twenty-four hours during the second day. Seven or eight readings on day 2 give sufficient data for students to accurately predict the growth of their dinosaur. Students should share the responsibility of measuring and recording the growth. Students might schedule

Students discover what the increase in length means in relation to the dinosaur's increase in area and volume

different members of their group to visit the classroom at one-hour intervals, perhaps at the change of class, to make the measurements. Contact with administration and colleagues might help this part of the activity go smoothly.

The last task for day 1, after students have taken and recorded the measurements of their dinosaur, is to work in small groups to complete sheet 4. The purpose of sheet 4 is to enable students to work with squares to discover the quadratic relationship between side lengths and areas of similar figures. The last question on the sheet asks students about the side-area relationship for two similar rectangles.

Day 2

Students should immerse their dinosaur in water as early in the school day as possible so that they can obtain as many data readings as feasible. Each dinosaur should be placed in a sealable plastic bag that has been filled with water to one-third its volume. The sealable plastic bag will cut down on the water spills. Students can easily take linear measurements with their dinosaur in the bag, or they can remove the dinosaur to gather the data. In either situation, they might spill some water from the plastic bag, so the area where students do the measuring should be "waterproofed."

After students have made their measurement recordings during class, they can work in small groups on sheet 5. Sheet 5 helps students discover the cubic relationship between edge lengths and volumes of cubes. Students who have not had any previous experience with drawing three-dimensional figures may need help sketching the cubes. Although the sketches are not a crucial aspect of this sheet, you may choose to show students how to draw correct sketches of cubes before beginning sheet 5. The last question on sheet 5 asks students about the side-volume relationship between similar solids.

Day 3

Students should complete the remaining questions on the first three worksheets. Students need to pool the measurement data that they have collected and then use the data to complete their individual graphs of linear growth. After students have shared their data, they should be able to predict the final length of their dinosaurs. Students can then verify their predictions by measuring the length of their grown dinosaurs. Students should use the initial measured length and the final measured length to obtain the scale factor between the original dinosaur and the grown dinosaur.

Using Activities from the *Mathematics Teacher* to Support *Principles and Standards*

This scale factor should then be used to predict the final cross-section area and the volume of the grown dinosaur. After making a calculated prediction for the area and volume of the grown dinosaur on the basis of the scale factor, students can check their predictions by measuring the area and volume of their dinosaurs. They can measure the area of the grown dinosaur by tracing the outline on sheet 2. To measure the volume of a fully grown dinosaur, students will need a container large enough to hold the dinosaur and a dish or tray as an overflow catcher. First, students should fill the container to the brim with water. When the dinosaur is carefully placed in the filled container, the water will overflow into the dish or tray underneath. The overflow of water is equivalent to the volume of the dinosaur. The volume of the overflow water collected in the dish or tray can then be measured by pouring it into the graduated cylinder.

The last item on sheet 3 asks students to design a poster that promotes the virtues of their growable dinosaurs. The dinosaurs generally increase their length by a factor of 6, or 600 percent. The result would be a 6^2, or 3600 percent, increase in area and a whopping 6^3, or 21600 percent, increase in volume.

When students understand the different proportional relationships among lengths, areas, and volumes of similar figures, they can appreciate why a giant like Paul Bunyan could not exist. For ease of calculations, suppose that Paul Bunyan is ten times the height of an average lumberjack. If the lumberjack is six feet tall, then Paul Bunyan would be sixty feet tall. His surface area would be 10^2, or 100, times as great as that of the lumberjack, and his volume would be 10^3, or 1000, times as great.

Would Paul Bunyan's bones be strong enough to support a thousandfold growth in volume or weight? Intuitively one might think that such a result is possible, since the bones of a giant would also increase in volume by a factor of 1000. However, the cross sections of the bones, not the volume, bear the weight of a body. The area of the cross section of a bone accounts for its strength, as Galileo noted. The cross section of a bone grows in a proportion identical to the growth of area. Paul Bunyan's area is larger than the area of the lumberjack by a factor of 100, but Paul Bunyan's weight is larger by a factor of 1000. Paul Bunyan's bones must therefore carry ten times their normal weight. The average human bone can support only about six times the normal weight that it usually carries. Thus, Paul Bunyan could not stand unless

a change occurred in his body structure, such as shorter, thicker legs like a hippopotamus or elephant. Evidence of the need for broader and more massive bones in larger animals is demonstrated by the fact that bones make up only 7 percent of the mass of a mouse but 18 percent of the mass of an average human (Thompson 1961).

Why can a mouse fall from a height of twenty feet and scurry away unhurt? A human falling from a proportional height would be seriously —if not fatally—injured. The answer lies again in the different proportional changes in length, area, and volume. By using a 6-foot-tall lumberjack and a mouse that is 0.6 feet long, some comparisons can be made. The mouse's length is only 1/10 the length of the man. The mouse's surface area is $(1/10)^2$, or 1/100, the surface area of the lumberjack; and its volume, or weight, is $(1/10)^3$, or 1/1000, the weight of the lumberjack. Thus, the weight of a mouse is reduced to 1/1000 that of the lumberjack, but its surface area is reduced to only 1/100 the surface area of the lumberjack. A falling mouse therefore has ten times the air resistance on its body as a falling lumberjack. The increased air resistance means that the mouse can survive a fall that would kill any large mammal (Haldane 1956).

A final application of the laws of mathematics imposed by nature can be seen in the types of mammals that live near the Arctic. In colder climates, the necessity of eating enough food to maintain body temperature is paramount. Body heat produced is a function of body mass or volume. Body heat lost is a function of exposed surface or surface area. A mammal the size of a lumberjack can survive in the Arctic, whereas the mouse cannot. The mouse has a reduced surface area that is 1/100 the man's surface area. However, the volume of the mouse, the ability to generate body heat, is only 1/1000 that of the lumberjack. The mouse must work ten times harder to maintain body heat. For that reason, mice or small mammals cannot live in extremely cold climates. As it is, a mouse consumes about one-fourth its weight each day, whereas the average human consumes about one-fiftieth his or her weight. Even in warm climates, a small mouse or shrew represents the lower limit of mammal size. A mammal much smaller than a mouse is an impossibility because the mammal would have to spend all its time eating to maintain body heat. These conclusions are summed up in Bergmann's law, which suggests a tendency for smaller animals to live nearer the equator and larger animals to live closer to the poles because of the ratio of their surface area to their volume.

Would Paul Bunyan's bones be strong enough to support a thousandfold growth in volume or weight?

Finally, students might also now understand why the main character in Gulliver's Travels had to be set free by his Lilliputian captors. According to the Jonathan Swift novel, Gulliver was twelve times the height of the average Lilliputian. Food intake is related to volume. Gulliver required 12^3, or 1728, times the amount of food that the average Lilliputian needed. At that rate, Gulliver would have bankrupted the economy of Lilliput in a matter of days. No wonder they set him free.

RESOURCES

Grow Dino, from Micro Mole Scientific; (509) 545-4904. 1–30 dinosaurs, $1.25 each, delivery included; 31 or more dinosaurs, $0.80 each, delivery included. Other growable figures can be used for this activity. However, their growth rates may differ from that of the Grow Dino described in this activity.

SOLUTIONS

Sheet 4

1.

Side ratios	Area ratios
1:2	1:4
1:3	1:9
1:4	1:16
1:5	1:25

2. The area ratio is the square of the side ratio.
3. 1:4   9:25   4:25   16:9
4. The amount of poster board should be increased by a factor of 4, or 2^2.

Sheet 5

1. Number of cubes

    1   8   27   64

2.

Side ratio	Volume ratio
1:2	1:8
1:3	1:27
1:4	1:64

3. The ratio of the volumes is the cube of the ratio of the sides.
4. 8:27   27:64   1:125
5. If you double the edges of a box, then the volume will be eight times as large, not twice as large.

References

Behr, Merlyn J., Richard Lesh, Thomas R. Post, and Edward Silver. "Rational-Number Concepts." In *Acquisition of Mathematics Concepts and Processes*, edited by Richard Lesh and Martha Landau. New York: Academic Press, 1983.

Cramer, Kathleen, Thomas Post, and Sarah Currier. "Learning and Teaching Ratio and Proportion: Research Implications." In *Research Ideas for the Classroom: Middle Grades Mathematics*, edited by Douglas Owen, 159–78. New York: Macmillan Publishing Co., 1993.

Haldane, J. B. S. "On Being the Right Size." In *The World of Mathematics*, vol. 2, edited by James R. Newman, 952–57. New York: Simon & Schuster, 1956.

Hart, Kathleen. "Ratio and Proportion." In *Number Concepts and Operations in the Middle Grades*, edited by James Hiebert and Merlyn Behr, 198–219. Reston Va.: National Council of Teachers of Mathematics, 1988.

Kaput, James, and Mary Maxwell West. "Missing Values Proportional Reasoning Problems: Factors Affecting Informal Reasoning Patterns." In *Development of Multiplicative Reasoning in the Learning of Mathematics*, edited by Guershon Harel and Jere Confrey, 237–91. Albany, N.Y.: State University of New York Press, 1994.

Lawton, Carol A. "Contextual Factors Affecting Errors in Proportional Reasoning." *Journal for Research in Mathematics Education* 24 (November 1993): 460–66.

Lesh, Richard, Thomas Post, and Merlyn Behr. "Proportional Reasoning." In *Number Concepts and Operations in the Middle Grades*, edited by James Hiebert and Merlyn Behr, 93–118. Reston, Va.: National Council of Teachers of Mathematics, 1988.

Piaget, John, Bärbel Inhelder, and Alina Szeminska. *The Child's Conception of Geometry*. New York: Basic Books, 1960.

Post, Thomas R., Merlyn J. Behr, and Richard Lesh. "Proportionality and the Development of Prealgebraic Understandings." In *The Ideas of Algebra, K–12*, 1988 Yearbook of the National Council of Teachers of Mathematics (NCTM), edited by Arthur F. Coxford, 78–89. Reston, Va.: NCTM, 1988.

Swift, Jonathan. *Gulliver's Travels*. New York: Grosset & Dunlap, 1947.

Thompson, D'Arcy. *On Growth and Form*. Cambridge, England: Cambridge University Press, 1961.

Editorial Comment: The following extensions were suggested by a reader and the author:

We expanded the activity by graphing the length versus time, in hours, for each color on the same graph. We then discussed how we could determine from the graph which dinosaur grew the fastest—a perfect opportunity to reemphasize slope. We also discussed why certain colors seemed to grow faster than others.

Advanced students could do a regression analysis to find which curve best fits the dinosaur's growth.
Deirdre McConnell, Hazelwood Central High School, Florissant, Missouri

A further extension might be to determine whether the dinosaurs that grew more quickly will also shrink more quickly back to their original size. Students could also explore how the "shrink curve" compares with the growth curve for their dinosaur. Since the shrinking takes a number of days, students might also try to predict the final "shrink" day.
Art Johnson

Dinosaur length:

Hour 0 _____   Hour 3 _____   Hour 6 _____

Hour 1 _____   Hour 4 _____   Hour 7 _____

Hour 2 _____   Hour 5 _____   Hour 8 _____

Graph of Dinosaur Growth

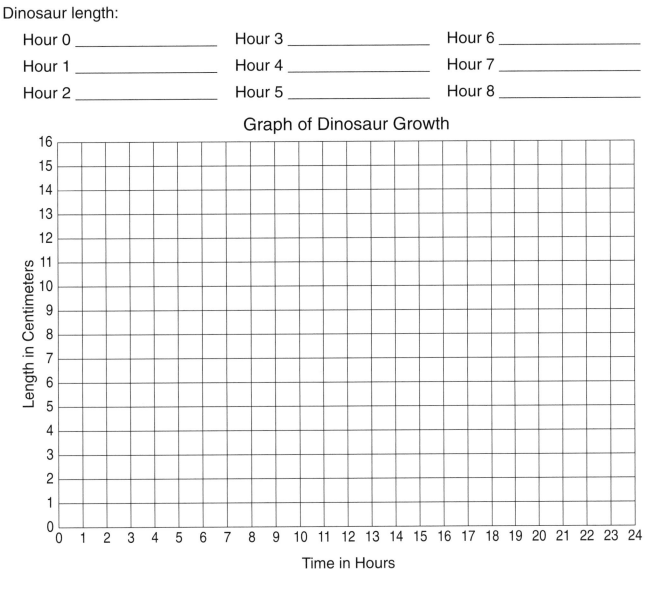

Time in Hours

1. Day 1: Length of original dinosaur   _____

2. Day 1: Description of the measured length (head to toe, wingspan, or other) _____

   _____

3. Day 3: Predicted length of the grown dinosaur on the basis of the graphed data _____

   _____

4. Day 3: Explanation of your prediction:

5. Day 3: Measured length of the grown dinosaur _____

6. $\dfrac{\text{Length of original dinosaur}}{\text{Length of grown dinosaur}}$ = _____

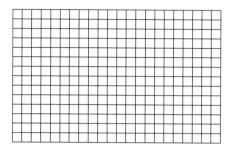

Outline of Original Dinosaur

1. Day 1: Area (cross section) of original dinosaur: _____ square units

2. Day 3: Predicted area of grown dinosaur on the basis of data from sheet 1: _____ square units

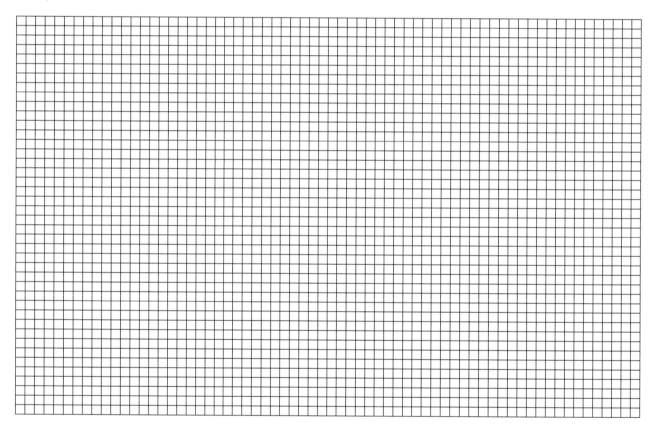

Outline of Grown Dinosaur

3. Day 3: Measured area of grown dinosaur: _____ square units

4. $\dfrac{\text{Area of original dinosaur}}{\text{Area of grown dinosaur}} = $ _____

1. Day 1: Measured volume of original dinosaur _____

2. Day 3: Predicted volume of grown dinosaur _____

3. Day 3: Measured volume of grown dinosaur _____

4. $\dfrac{\text{Length of original dinosaur}}{\text{Length of grown dinosaur}} = $ _____

$\dfrac{\text{Area of original dinosaur}}{\text{Area of grown dinosaur}} = $ _____

$\dfrac{\text{Volume of original dinosaur}}{\text{Volume of grown dinosaur}} = $ _____

5. Day 3: How are the three ratios related?

6. Day 3: Do the ratios fit the expected relationship? Explain your answer.

7. Day 3: Design a poster that promotes these growable dinosaurs. As part of your poster, be sure to compare the size of the grown dinosaur with that of the original.

For this activity, you will draw some squares on the grid below. The dimensions of each square are shown above the grid. Draw the square below its dimension. For example, a 2 × 2 square is a square with sides that are two units long.

Square 1	Square 2	Square 3	Square 4	Square 5
1 × 1	2 × 2	3 × 3	4 × 4	5 × 5

1. Use the results of your sketched squares to fill in the data below.

Side ratios   Area ratios

$$\frac{\text{Side square 1}}{\text{Side square 2}} = \underline{\hspace{3cm}}$$ $$\frac{\text{Area square 1}}{\text{Area square 2}} = \underline{\hspace{3cm}}$$

$$\frac{\text{Side square 1}}{\text{Side square 3}} = \underline{\hspace{3cm}}$$ $$\frac{\text{Area square 1}}{\text{Area square 3}} = \underline{\hspace{3cm}}$$

$$\frac{\text{Side square 1}}{\text{Side square 4}} = \underline{\hspace{3cm}}$$ $$\frac{\text{Area square 1}}{\text{Area square 4}} = \underline{\hspace{3cm}}$$

$$\frac{\text{Side square 1}}{\text{Side square 5}} = \underline{\hspace{3cm}}$$ $$\frac{\text{Area square 1}}{\text{Area square 5}} = \underline{\hspace{3cm}}$$

2. What pattern can you see between the side ratio and the area ratio for these pairs of squares?

3. Use this pattern to predict the area ratio between the following pairs of squares:

$$\frac{\text{Area square 2}}{\text{Area square 4}} = \underline{\hspace{3cm}}$$ $$\frac{\text{Area square 3}}{\text{Area square 5}} = \underline{\hspace{3cm}}$$

$$\frac{\text{Area square 2}}{\text{Area square 5}} = \underline{\hspace{3cm}}$$ $$\frac{\text{Area square 4}}{\text{Area square 3}} = \underline{\hspace{3cm}}$$

4. Use what you have learned to solve the following problem:

The owner of a poster company has decided to double the side lengths of all its posters. The owner tells you to double the amount of poster board to account for this increase. What would you tell her?

From the *Mathematics Teacher*, February 2000

1. Use centimeter cubes to build the following cubes. For example, a $3 \times 3 \times 3$ cube is a cube that has edges that are three centimeters long. Record how many centimeter cubes you need to build each cube. Make a sketch of the cube under the recorded number of cubes.

Cube 1	Cube 2	Cube 3	Cube 4
$1 \times 1 \times 1$	$2 \times 2 \times 2$	$3 \times 3 \times 3$	$4 \times 4 \times 4$

Number of
cubes needed

2. Use the data that you collected to fill in the following blanks.

Side ratios   Volume ratios

$$\frac{\text{Side cube 1}}{\text{Side cube 2}} = \underline{\hspace{3cm}}$$ $$\frac{\text{Volume cube 1}}{\text{Volume cube 2}} = \underline{\hspace{3cm}}$$

$$\frac{\text{Side cube 1}}{\text{Side cube 3}} = \underline{\hspace{3cm}}$$ $$\frac{\text{Volume cube 1}}{\text{Volume cube 3}} = \underline{\hspace{3cm}}$$

$$\frac{\text{Side cube 1}}{\text{Side cube 4}} = \underline{\hspace{3cm}}$$ $$\frac{\text{Volume cube 1}}{\text{Volume cube 4}} = \underline{\hspace{3cm}}$$

3. What pattern can you see between the side ratio and the volume ratio for these pairs of cubes?

4. Use this pattern to predict the volume ratio between the following pairs of cubes:

$$\frac{\text{Volume cube 2}}{\text{Volume cube 3}} = \underline{\hspace{3cm}}$$ $$\frac{\text{Volume cube 3}}{\text{Volume cube 4}} = \underline{\hspace{3cm}}$$

Predict the volume ratio between a 1×1 cube and a 5×5 cube.

5. Use the pattern information that you have learned to solve the following problem:

The president of a candy factory has decided to double the number of chocolates in his boxes of candy. He tells you that you need to design a new box that has edges that are twice as long as those of the original box. What would you tell him?

The Volume of a Pyramid: Low-Tech and High-Tech Approaches

Masha Albrecht

January 2001

Edited by A. Darien Lauten, *Rivier College, Nashua, New Hampsire*

> **I used this lesson with students who had no previous spreadsheet experience.**

This lesson came about spontaneously during a geometry unit on volume. I had used the lesson shown here in activity sheet 1, in which students use cubic blocks to rediscover the formulas for volumes of right prisms, that is, $V = Bh$ and $V = lwh$. This lesson was a simple review for my tenth-grade class, and they completed it easily before the end of the period. With the wooden cubes still on their desks, most of them used the remaining time to build towers and other objects. I noticed that many students piled the cubes into bumpy pyramidal shapes. Because the next day's lesson involved studying the volume of pyramids, I wondered whether these bumpy shapes could be useful for discovering the volume of a real pyramid with smooth sides. Students could compare the volumes of these "pyramids of cubes" with the volumes of corresponding right prisms and perhaps discover the ratio 1/3 to obtain the formula for the volume of a pyramid, $V = (1/3)Bh$. As it turns out, the ratio of 1/3 does not become evident right away. To my students' delight, we found that using a spreadsheet is an excellent way to investigate this problem. My geometry classes had not used spreadsheets before, and the students enjoyed the experience of using the efficiency of technology to compare hundreds—and even thousands—of shapes with ease.

Prerequisites: Students with only very basic mathematical knowledge can benefit from this lesson. Students should have some skill at describing a pattern with an algebraic equation and some familiarity with a spreadsheet. However, I used this lesson with students who had no previous spreadsheet experience.

Grade levels: Although I originally used this lesson with a regular tenth-grade geometry class, the lesson is appropriate for students at different levels and with different abilities. A prealgebra class could do the low-tech part of the lesson, in which students find patterns by using blocks, but they would need help with the formulas for the spreadsheet. Eleventh-grade or twelfth-grade students with more advanced algebra skills could be left on their own to find the spreadsheet formulas and could be given the difficult challenge of finding the closed formula for the volume in the "pyramid of cubes" column on activity sheet 2. A calculus class could find the limit of the ratio column as n goes to infinity before they check this limit on the spreadsheet.

Materials: The entire lesson works well in a two-hour block or in two successive fifty-minute lessons, with the low-tech lesson in the first hour and the high-tech spreadsheet lesson in the second. Cubic blocks are needed for the low-tech lesson. Because approximately forty blocks are needed for each group of four students, large classes will need many blocks. If you do not have enough blocks, groups can share. Simple wooden blocks work best; plastic linking cubes do not work as well, because their extruding joints can get in the way when students build the pyramids.

Spreadsheet software is needed for the high-tech lesson. If you are using a separate computer lab, sign out the lab for the second hour of this activity.

For the low-tech extension lesson, the following additional materials are needed: a hollow pyramid and prism with congruent bases and heights, as well as water, sand, rice, or small pasta.

TEACHING SUGGESTIONS

Sheet 1: Using cubic blocks—volume of prisms

This activity sheet is elementary, and more advanced students can skip it. Have students work in groups, with one set of blocks per group. Often one student quickly sees the answers without needing manipulatives, but the other group members are too shy to admit that they need to build the

---

Masha Albrecht teaches at the Galileo Academy of Science and Technology, her neighborhood school in San Francisco. She is interested in finding appropriate uses of technology for mathematics learning.

shapes. Require that each group build most of the solids, even if students protest that this activity seems easy.

Sheet 2: Using cubic blocks—volume of pyramids

Students may initially have difficulty understanding what the "pyramids of cubes" look like. Make sure that they build the one with side length 3 correctly. After using the blocks to build a few of the shapes, students recognize the patterns and start filling in the table without using the blocks. Calculating decimal answers for the last column of ratios instead of leaving answers in fraction form helps students look for patterns. Have a whole-class discussion about questions 4, 5, and 6 after students have had a chance to answer these questions in smaller groups, but do not reveal the answers to these questions. Students discover the answers when they continue the table on the spreadsheet.

The last row of the table, where students generalize the results for side length n, is optional. On the spreadsheet, students do not need the difficult closed formula for the second column. They can instead use the recursive formula, which is easier and more intuitive. The solutions include more explanation.

Sheet 3: Using a spreadsheet—volume of pyramids

This activity sheet is designed for students who have some spreadsheet knowledge. Having one pair of students work at each computer is useful if at least one student in each pair knows how to use computers and spreadsheets. For students who have no experience with spreadsheets, you can use this activity sheet as the basis for a whole-class discussion while demonstrating the process on an overhead-projection device. Do not bother photocopying activity sheet 3 for students who are familiar with spreadsheets. Instead ask them to continue the table from activity sheet 2, and give them verbal directions as needed.

Group members may be too shy to admit that they need to build the shapes

SOLUTIONS

Sheet 1, part 1

1.

Length	Width	Height	Volume
2 units	2 units	4 units	**16 cubic units**
1 unit	2 units	3 units	**6 cubic units**
2 units	2 units	**2 units**	8 cubic units
0.5 units	2 units	2 units	**2 cubic units**

2. $V = lwh$

Sheet 1, part 2

1.

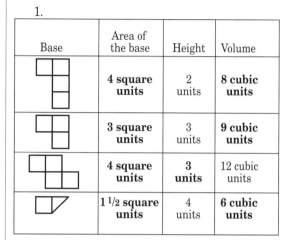

Base	Area of the base	Height	Volume
	4 square units	2 units	**8 cubic units**
	3 square units	3 units	**9 cubic units**
	4 square units	3 **units**	12 cubic units
	1 ½ square units	4 units	**6 cubic units**

2. $V = Bh$, where B is the area of the base.

Sheet 2

1. 8 cubic units
2. 5 cubic units
3.

Length of Side	Volume of Cubic Solid	Volume of "Pyramid of Cubes"	Volume of "Pyramid" Divided by Volume of Cubic Solid
1	1	1	1
2	8	5	5/8 = 0.625
3	27	14	14/27 ≈ 0.518
4	64	30	30/64 ≈ 0.469
5	125	55	55/125 = 0.440
6	216	91	91/216 ≈ 0.421
7	343	140	140/343 ≈ 0.408
8	512	204	204/512 ≈ 0.398
9	729	285	285/729 ≈ 0.391
10	1000	385	385/1000 = 0.385
n (if you can)	n^3	Volume of **previous n** + n^2 or $(1/3)n^3$ + $(1/2)n^2$ + $(1/6)n = (n/6) \cdot (n+1)(2n+1)$	$[(n/6)(n+1) \cdot (2n+1)]/n^3$

Some students may be interested in a derivation of the closed formula in the last cell of the "pyramid of cubes" column. One way to derive the formula from the information in the chart is to begin by establishing that the formula is a cubic function. Students who are familiar with the method of finite differences can see that the relationship is a cubic because the differences become constant after three iterations.

```
  1
     \  4
  5  /     \  5
     \  9  /     \  2
 14  /     \  7  /
     \ 16  /     \  2
 30  /     \  9  /
     \ 25  /     \  2
 55  /     \ 11  /
     \ 36  /     \  2
 91  /     \ 13  /
     \ 49  /     \  2
140  /
```

When students know that the formula is a cubic, they know that they can write it in the form $f(n) = an^3 + bn^2 + cn + d$, where n is the side length. Because the four constants a, b, c, and d are unknown, they can be treated as variables for now. Students can use the first four rows of the data in the table to see that $f(1) = 1$, $f(2) = 5$, $f(3) = 14$, and $f(4) = 30$. They can write the following system of equations:

$$a(1)^3 + b(1)^2 + c(1) + d = 1$$
$$a(2)^3 + b(2)^2 + c(2) + d = 5$$
$$a(3)^3 + b(3)^2 + c(3) + d = 14$$
$$a(4)^3 + b(4)^2 + c(4) + d = 30$$

That system is equivalent to the following system:

$$a + b + c + d = 1$$
$$8a + 4b + 2c + d = 5$$
$$27a + 9b + 3c + d = 14$$
$$64a + 16b + 4c + d = 30$$

However students solve this system, they find that $a = 1/3$, $b = 1/2$, $c = 1/6$, and $d = 0$, from which students can obtain the formula shown in the chart. The system is actually not very difficult to solve by hand using linear combinations.

4. Accept any reasonable answer at this point. Such answers might be similar to, "The ratio gets smaller as the shapes get bigger." In fact, the ratio in the last column approaches 1/3, or 0.33333. . . .

5. Again, accept any reasonable answer. The ratio approaches 1/3 because the pyramid of cubes becomes a closer approximation of an actual smooth-sided pyramid. The size of the cubic blocks does not change as the pyramids become larger, so bumps created by the edges of the blocks are less significant as the "pyramid" becomes larger. If students are familiar with the notion of a limit, you can discuss how the limit of these larger and larger shapes is an infinitely large pyramid with completely smooth sides.

6. Although the ratio in the last column keeps getting smaller, it never reaches 0. Let

students discuss this result, but do not reveal the answer.

Sheet 3

1. and 2.

Length of Side	Volume of Cube	Volume of "Pyramid"	Volume of "Pyramid" divided by Volume of Cube
1	1	1	
2	8	5	

3. Although the formulas are displayed here, numbers should show in the cells on the students' spreadsheets.

	A	B	C	D
1	Length of Side	Volume of Cube	Volume of "Pyramid"	Volume of "Pyramid" Divided by Volume of Cube
2	1	1	1	=C2/B2
3	2	8	5	

4.

	A	B	C	D
1	Length of Side	Volume of Cube	Volume of "Pyramid"	Volume of "Pyramid" Divided by Volume of Cube
2	1	1	1	=C2/B2
3	2	8	5	=C3/B3

5.

	A	B	C	D
1	Length of Side	Volume of Cube	Volume of "Pyramid"	Volume of "Pyramid" Divided by Volume of Cube
2	1	1	1	=C2/B2
3	2	8	5	=C3/B3
4	=A3+1	=A4^3	=C3+A4^2	=C4/B4

6. A few sample rows are shown here.

Length of Side	Volume of Cube	Volume of "Pyramid"	Volume of "Pyramid Divided by Volume of Cube
196	7 529 536	2 529 086	0.335888692
197	7 645 373	2 567 895	0.335875699
198	7 762 392	2 607 099	0.335862837
199	7 880 599	2 646 700	0.335850105
200	8 000 000	2 686 700	0.3358375

Bumps created by the edges of the blocks become less significant as the "pyramid" grows larger

The Volume of a Pyramid

7. How students create this graph varies depending on the spreadsheet software and the platform. To select the side-length column and the nonadjacent ratio column, first select one column, then select the other while holding down the control key. Excel users should look for the Chart Wizard icon on the menu bar, click on this icon after selecting the side length and ratio column, and follow the menu choices until the appropriate graph appears.

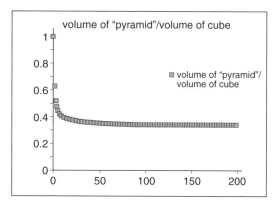

18. The numbers in the last column get closer and closer to 1/3.
19. No. The ratio will always be higher than 1/3.
10. $V = (1/3)Bh$.

Possible extensions

My students enjoyed moving away from the computers for this low-tech finale. If you have a hollow pyramid-and-prism set that has congruent bases and congruent heights, have students use the pyramid as a measuring device to fill the prism with water, sand, rice, or pasta. They should find that three pyramids of water or sand fill the prism exactly to the brim.

I ended the lesson by giving students a picture of some Egyptian pyramids from a book on architecture. The caption to the picture includes measurements, so students can calculate the volume of one of the actual pyramids.

REFERENCE

Norwich, John Julius. *World Atlas of Architecture*. New York: Crescent Books, 1984.

The pyramid of Cheops, the biggest of the three pyramids at Giza, measures 230.5 meters (756 feet) at its base and is 146 meters high. The slope is 51° 52'. At the center is the pyramid of Chephren. Although it is 215 meters (705 feet) at its base and 143 meters (470 feet) high, it appears higher because of its steeper slope (52° 20'). The pyramid of Mycerinus, in the foreground, is the smallest of the three. It measures 208 meters (354 feet) at its base and 62 meters (203 feet) in height, with a slope of 51°.

Using Activities from the *Mathematics Teacher* to Support *Principles and Standards*

Part 1: Volume of a rectangular box

   1. Construct each solid with your cubic blocks, and complete the chart. Use your imagination for the last answer.

Length	Width	Height	Volume
2 units	2 units	4 units	
1 unit	2 units	3 units	
2 units	2 units		8 cubic units
0.5 units	2 units	2 units	

   2. Write a formula for the volume of a rectangular box. _____

Part 2: Volume of a right prism

   1. Construct each solid with your cubic blocks, and complete the chart. Use your imagination for the last answer.

Base	Area of Base	Height	Volume
		2 units	
		3 units	
			12 cubic units
		4 units	

   2.   Write a formula for the volume of any right prism. _____

Although we cannot build exact pyramids with cubes, we can approximate them by building "pyramids of cubes" such as the two pictured below. You will compare the volume of a "pyramid of cubes" with the volume of the prism having the same base and height.

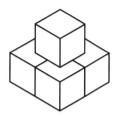

"Pyramid of cubes" with a
square base of side length 2
and height of 2

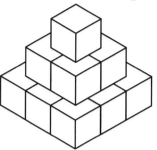

"Pyramid of cubes" with a
square base of side length 3
and height of 3

1. Find the volume of a cubic solid with a side of length 2. _____

2. Find the volume of the "pyramid of cubes" with a square base of side length 2 and a height of 2 (pictured above)._____

3. Complete the chart below. In the last column, compute the ratio of the number in the third column divided by the number in the second column.

Length of Side	Volume of Cubic Solid	Volume of "Pyramid of Cubes"	Volume of "Pyramid" Divided by Volume of Cubic Solid
1			
2			
3			
4			
5			
6			
7			
8			
9			
10			
n (if you can!)			

4. What happens to the ratio in the last column as your solids become larger?

5. Why do you think that you obtain this result?

6. Does the ratio in the last column ever become 0?

As you can tell, finding the pattern in the last column of your table is difficult unless you continue the table. You can create a spreadsheet to do the work for you instead of doing the work by hand.

1. In a spreadsheet, type the headings for the four columns of your table, as shown. You may want to abbreviate the headings.

Length of Side	Volume of Cube	Volume of "Pyramid"	Volume of "Pyramid" Divided by Volume of Cube

2. Enter the values for the first two rows into your spreadsheet. Do not enter numbers for the last column, because you will use a formula to cause the spreadsheet to calculate these values.

Length of Side	Volume of Cube	Volume of "Pyramid"	Volume of "Pyramid" Divided by Volume of Cube
1	1	1	
2	8	5	

3. Enter a formula for ratio into the first empty cell in the last column. Remember that the formulas in a spreadsheet begin with an "=." Do not just type in the number 1.

4. Copy the ratio formula that you just wrote into the cell below it. Your spreadsheet should look something like the following:

Length of Side	Volume of Cube	Volume of "Pyramid"	Volume of "Pyramid" Divided by Volume of Cube
1	1	1	1
2	8	5	0.625

5. The next row of your spreadsheet will contain only formulas. Enter all four appropriate formulas for the next row. For help, use the patterns that you noticed when you built the shapes with blocks. You can also work with other students.

6. Select the row of formulas that you just created, and copy them into the next row. Continue to copy down into more and more rows. Use any shortcut that your software allows, such as Fill Down, until your table is long enough that you are sure of a pattern in the last column.

7. Use the graphing feature of your spreadsheet to make a graph of the ratio numbers in the last column.

Use your spreadsheet to answer the following questions. Some of them are repeated from sheet 2.

8. What happens to the ratio in the last column as the solids become larger?

9. Will the ratio in this column ever be 0? Why or why not?

10. You can use your experience with the "bumpy" pyramids that you made with blocks to generalize the outcome for any pyramid. If a pyramid has a base of area B and a height of h, write a formula for its volume.

A S(t)imulating Study of Map Projections: An Exploration Integrating Mathematics and Social Studies

Jesse L. M. Wilkins and David Hicks

November 2001

Edited by Betty Krist, *University at Buffalo, Buffalo, New York*

> *Students hold major misconceptions about proportions, locations, and perspective*

As technological advances continue to help more people make connections with the entire world, students must understand how to use and interpret information shown in different maps of the world (Geography Education Standards Project 1994; Freese 1997). However, mental-mapping research suggests that students in the United States have major misconceptions about proportions, locations, and perspective when they work with maps (Dulli and Goodman 1994; Stoltman 1991).

Studying different map projections offers a unique opportunity to integrate mathematics and geography. In fact, the National Council of Teachers of Mathematics (NCTM 1989) emphasizes the importance of making connections between mathematical topics and such other content areas as social studies. Furthermore, the National Council for the Social Studies (NCSS 1994) emphasizes such representations of the earth as maps and globes. This emphasis includes using, creating, and interpreting data related to different map projections. The activities in this article allow students to investigate the distortions associated with different map projections and encourage them to discuss the assumptions that the projections may perpetuate about the world. Further, the activities provide an interesting real-world context for discussing methods of simulation for collecting data and making estimates, as well as for plotting points in a Cartesian plane.

Maps have an enormous influence on our perception of the world. The Mercator projection, shown in figure 1, with either a European- or American-centered representation of the world, is one of the most recognizable projections of the earth (Snyder 1987). Figure 2 shows the Robinson map, with its nonrectangular shape, which attempts to present a more accurate representation of the world. The National Geographic Society used the Robinson projection as its official view of the world from 1988 to 1998 (Makower 1992). The Mollweide projection, shown in figure 3, is an equal-area map within an ellipse. It maintains the proportional relationships related to area. However, these projections are only three of the many that exist. For centuries, as far back as the Greek mapmaker and astronomer Ptolemy, articles and books have focused on the problem of representing the spherical world on a flat surface (Snyder 1987). No one projection can be deemed the "best" representation of the globe because the way that information is transformed from the spherical earth to a useful flat map requires that cartographers select a projection that best fits their own goals and objectives (Wikle 1991).

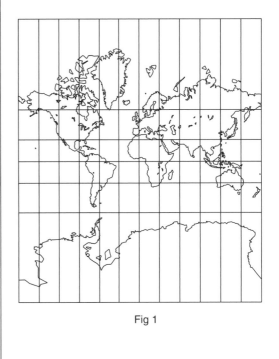

Fig 1

Jesse (Jay) Wilkins and David Hicks both teach at Virginia Polytechnic Institute and State University, Blacksburg, Virginia. Wilkins teaches mathematics education, and his areas of research include educational opportunity and quantitative literacy. Hicks teaches social studies education, and his interests include citizenship, technology, and interdisciplinary connections.

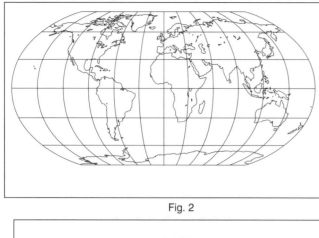

Fig. 2

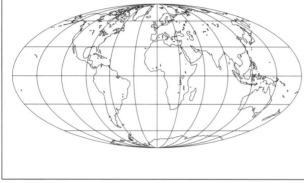

Fig. 3

Map projections themselves are based on the needs of a given situation and on the characteristics that are being investigated. Most map projections are constructed to maintain area, distance, direction, shape, or a combination of these characteristics. Some mapmakers are interested in showing accurate relationships of the areas of land and water masses. Other maps are concerned with maintaining accurate representations of distances between points on the surface of the earth. Others maintain accurate direction in terms of north, south, east, and west, whereas others try to make the shape of land and water areas accurately represent the true shapes found on the earth's surface. Although many people assume that maps represent reality, accurately representing all the desirable properties within one projection is not possible. All flat maps distort one or more aspects of their representation of the earth. Therefore, the mapmaker must decide the characteristic that is to be shown accurately at the expense of the others or create a compromise among several properties (Committee on Map Projections of the American Cartographic Association 1986; Snyder 1987).

Although the mathematics underlying the maintenance of these properties for different map projections may be beyond the level of many mid-dle and high school students, the interpretations and social implications that may be associated with them are not. The activities presented in this article combine mathematics and geography through investigating the proportion of land and water that covers the earth. The activities serve as a point of entry through which students can become familiar with characteristics of different projections or representations of our world while they hone their estimation and graphing skills in an interesting context. In addition, the activities encourage students to investigate the concepts of expected value and the power of simulation to make estimates that are otherwise almost impossible to make.

TEACHER'S GUIDE

Grade levels: 7–12

Objectives: The main objective of the activities that follow is to allow students to investigate the ways that different map projections distort various characteristics of the earth. Specifically, students use simulation methods to estimate the percent of the earth that is covered by water on the basis of four different representations of the earth. In addition, students practice graphing in a coordinate plane and hone their estimation skills through simulation. Students are also encouraged to think critically about how different map projections may affect people's world views with regard to their relationships with other people and countries.

Prerequisites: These activities presume some familiarity with graphing points on a Cartesian plane. Some knowledge of random sampling and Monte Carlo simulation might be helpful; however, the activities themselves may introduce these topics. Sheet 1 gives additional explanation of the Monte Carlo method. See Sobol' (1994); Travers and others (1985); Stout, Marden, and Travers (1999); Wilkins (1999).

Materials: An unbreakable globe—an inflatable globe works best; activity sheets; map projections given in figures 1, 2, and 3, enlarged to at least 8.5 × 11 inches; graph-paper transparencies made from standard five-division-per-centimeter graph paper; graphing calculator; atlas

AN INTRODUCTORY ACTIVITY: USING A GLOBE TO ESTIMATE THE EARTH'S WATER COVERAGE

A globe is the most accurate representation of the earth with regard to area, distance, direction, and shape. Therefore, a measure of the water coverage presented on a globe gives the best estimate of the percent of the earth that is covered with water.

Goals

The goals of the activity discussed on sheets 1 and 2 are to investigate characteristics of the globe and to determine the percent of the earth's surface that is covered by water. This activity introduces simulation as a method of estimation and helps furnish evidence for the validity of the method. The results of this activity provide a benchmark for comparing estimates of water coverage on the basis of different map projections.

Procedure

Begin by asking your students whether they know or can guess the percent of the earth that is covered by water. Encourage students to discuss ways of estimating this percent. At this point, depending on the background of the students, you might discuss the ideas of expected value and simulation. For example, you might discuss what they would expect if they flipped a penny many times and how they can use such expectations to make estimates.

To prepare for the simulation, have each student make an ink dot on the end of his or her index finger. You might discuss this procedure and explain that you designate the index finger for recording the data to maintain consistency. The students should then toss the globe—preferably an inflatable globe—around the room. When students catch it, they should note whether the ink dot is touching water or land. They can record data on sheet 1. The class should decide on a prescribed number of tosses but should do at least one hundred tosses. On the basis of the data, have the students calculate the percent of times that their finger pointed to water compared to the total number of tosses. These calculations give an estimate for the percent of the earth that is covered by water.

Alternatively, if you have several globes, students can work in groups, with one person as a recorder. The data can then be pooled for the class, and more data points can be obtained in less time.

Expectations

Approximately 70 percent of the earth's surface is covered by water. The simulation activity should give a reasonably good estimate. However, instead of just presenting 70 percent as the "answer," you might encourage students to determine the actual percent on their own. For example, the land area of the earth is divided into seven continents. Teachers can tell students to use sheet 2 to calculate the total land area. If students assume that the earth is a sphere, they can use the diameter of the earth, or radius r, which can also be found in an atlas, to estimate the surface area of the earth (surface area of a sphere = $4\pi r2$). These calculations together can be used to calculate the percent of the earth that is covered by land and the percent that is covered by water.

Students should compare the results of the simulation with their own expectations, as well as with the true percent. A discussion of the simulation results should follow, with emphasis on the variability of estimates—for example, students might discuss possible results of repeating the simulation, the reasonableness of the estimates, and the power of simulation to provide reasonable estimates, that is, its proximity to 70 percent.

ESTIMATING WATER COVERAGE USING DIFFERENT MAP PROJECTIONS

Goal

The goal of this activity is to have students use three different map projections to estimate the percent of the earth's surface that is covered by water. Students create a coordinate system for each projection and plot randomly produced points on the different projections. They use these points to estimate the percent of the earth's surface that is covered with water. By comparing the estimates of each projection with the estimates produced by using the globe, students can make judgments about the effects of different projections on the representation of the earth.

Preparation

Three different map projections are given in figures 1–3. The maps can be used in the 8.5 × 11 size, or you can enlarge them further. Suggested graph paper is standard five-division-per-centimeter, which should be transferred to one transparency per group. If you enlarge the maps, you need to have a larger sheet of graph paper, but you should not enlarge the graph paper on a copy machine, since doing so changes the size of the grid.

Background

The three map projections in figures 1–3 were produced using MicroCam for Windows (Loomer 1999), which can be downloaded from wolf.its.ilstu.edu /microcam/. The three map projections selected for this activity each present a different perspective on maintaining area. However, many

The students should then toss the globe around the room

other projections could be produced using Micro-Cam in subsequent investigations. The following briefly describes the projection properties with regard to maintaining equality of area:

- Figure 1: Mercator projection—most commonly used, but it does not preserve area

- Figure 2: Robinson projection—a compromise map that distorts area, distance, direction, and shape to balance the error of projection properties

- Figure 3: Mollweide projection—an equal-area map, that is, it preserves area

Sheet 3a gives further details. A useful resource that provides other map projections and describes their properties can be found at www.colorado.edu /geography/gcraft/notes/mapproj/mapproj_f.html.

Procedures

To collect data efficiently, students should work in pairs or in small groups, with one person generating the numbers and another recording the location of the point. You can either give each group a different map or have each group investigate all the maps, depending on the amount of time available. Encourage students to think about why and whether estimates of water coverage might change when they use the different map projections. Sheet 3a gives additional information that can guide students in making conjectures about water coverage. Have the students think about how they might use the flat map projections to estimate the percent of water coverage. At some point, on the basis of the progress of the discussion, you might ask students to think about how they could transfer the procedures that they used for the globe simulation to the flat maps. They can discuss how to plot a random sample of points on the projections. Students may suggest throwing darts at the maps. The teacher should explain the nonrandom nature of that technique and indicate that the ability to aim may affect where the darts land. However, the dart-throwing technique is the essence of the technique to be used and serves as a good introduction to discussing how throwing darts could be made random, that is, by using a coordinate system and randomly choosing points on the grid.

To create a coordinate system for the maps, students overlay the graph-paper transparency onto the map projection to create a way of plotting points on the map. Students should label their graph-paper transparency with appropriate coordinate values. You might next want to discuss with students such methods of creating random num-

Students may suggest throwing darts at maps

bers as telephone directories, random-number tables, and calculators.

Several calculators have a randInt function that randomly generates a pair of integers within the assigned coordinate system. Calculators and spreadsheets with only a rand function can produce a random number between 0 and 1, inclusive. The combination Int(100*rand()) can be used to produce a random integer between 0 and 100, inclusive. A coordinate pair can be produced by using two randInt functions. For example, on the TI-83, enter the following functions within brackets and separated by a comma: {randInt(0,100), randInt(0,50)}. By continuing to press the ENTER button, you can continue to produce coordinate pairs. You may need to explain what students should do with values that fall outside the range of the nonrectangular maps, for example, the Mollweide and Robinson maps. Students should proceed by producing random coordinate pairs and determining whether their locations are in water or on land. As in the globe activity, students should do at least one hundred trials; they should keep track of their data on sheet 3b. However, creating a class estimate that combines estimates from all the groups might be more efficient. For example, if each pair of students in a class of twenty students plotted fifty points per map, a total of five hundred data points would be obtained. Also, by dividing the work, plotting the points does not become the major emphasis and time-consuming portion of the activity.

Students should compare the estimates of the percent of water coverage for each of the map projections. They next discuss their findings and suggest the reasons that they might have obtained these results, especially in comparison with the actual percent of water coverage. In addition, a comparison of the different estimates produced by each pair for a given map might lead to a discussion of precision and variability. Calculating an overall average of the different estimates furnishes a measure of center that can be used to discuss the distribution of estimates—for example, outliers and range—and may also lead to a discussion of the central limit theorem.

Expectations

This activity is exploratory, in that no answers are set. However, certain trends might be expected; possible explanations for these trends, as well as explanations of the conditions under which they occur, follow.

The Mercator map projection, shown in figure 1, greatly distorts areas as one gets further from

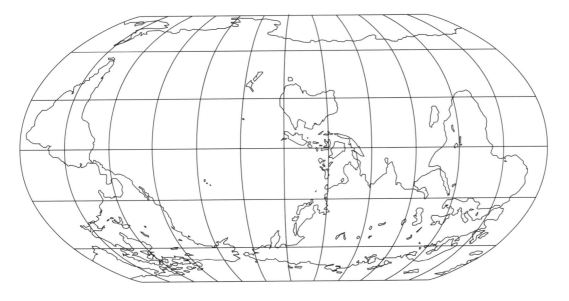

Fig. 4. An inverted Robinson projection presenting an Australian perspective, that is, with Australia centered and on top of the world.

the equator. In the southern part of the world, for example, Antarctica, the land masses appear much larger than they should be. Similarly, Greenland, in the north, appears to be almost three times its actual size. Thus, the Mercator projection exaggerates the amount of land area. Therefore, the percent of water covering the earth might be underestimated, even though the area of the Arctic Ocean is also exaggerated. The Robinson map is considered to be a compromise projection. Although it does not maintain equal area, it can furnish a reasonable estimate of water coverage because all the properties are compromised to reduce the overall error found in each property. The best estimate should come from the Mollweide map projection, since it maintains equal area in all regions of the map. Of course, no matter which map projection is used, all the estimates are subject to sampling error.

The structure of the Mercator map offers an interesting opportunity to evaluate sampling error. If the points tend to be located near the equator, the estimate of water coverage for that area may be quite good, since the Mercator map tends to maintain area reasonably well around the equator. However, if the points tend to be located toward the poles, the estimate may be very different from the expected result. In anticipation of this possibility, having students keep track of the latitude of the points would be an interesting extension. Students can use the y-coordinates of the graph paper to divide the map horizontally into thirds—that is, the northern third, the equatorial third, and the

southern third—and then record not only whether the point is located in water or on land but also the region in which it is located. See sheet 4. On the basis of the data obtained, students might be able to make conjectures about the source of some of the variability in different estimates.

In an additional extension that might be appropriate for a statistics class, students could test whether the estimates obtained for each projection differ significantly from the hypothesized 70 percent. For example, students might hypothesize that an estimate of water coverage that is based on the Mercator map would be less than 70 percent. Similarly, students might hypothesize that the estimate that used the Mollweide map would not differ from 70 percent. Students could test these hypotheses by using resampling techniques or another appropriate statistical test, for example, the Z-test.

CONCLUSIONS

Too often, time becomes the stumbling block in developing lessons that allow students to actively inquire and to make real-world connections. These activities furnish an example of the possibilities for ongoing collaboration between mathematics and social studies teachers, as well as for integration within each classroom. See sheet 5. The lesson recognizes the importance of giving students experiences that focus on their geographic and quantitative literacy needs as tomorrow's global citizens. The activities allow students to see maps as sources of data that can present very different

A S(t)imulating Study of Map Projections

images and perspectives of the world. Students can continue to explore different map projections, for example, investigating the shape and proportional size of such specific areas as Greenland, which appears as large as South America on many of the map projections found on school walls and in textbooks. Such activities can encourage students to reflect on the ways that certain perspectives, impressions, and attitudes form as part of their worldview with regard to their beliefs about other people and countries. Representing north as up is only a convention, but most Americans would think that a map with Australia in the "upper hemisphere," as shown in figure 4, was odd, although that representation is also valid. Such discussions open possibilities within issue-centered classrooms for students to examine how perspectives and attitudes are influenced by the uncritical acceptance of the many representations of people and the world, beyond map projections, that everyone comes into contact with on a daily basis.

REFERENCES

Committee on Map Projections of the American Cartographic Association. *Which Map Is Best?* Projections for World Maps. Falls Church, Va.: American Congress on Surveying and Mapping, 1986.

Dana, Peter H. "Map Projections." www.colorado.edu /geography/gcraft/notes/mapproj/mapproj_f.html. World Wide Web.

Dulli, Robert E., and James M. Goodman. "Geography in a Changing World: Reform and Renewal." *NASSP Bulletin* 78 (October 1994): 19–24.

Freese, J. R. "Using the National Geography Standards to Integrate Children's Social Studies." *Social Studies and the Young Learner* 10 (November–December 1997): 22–24.

Geography Education Standards Project. *Geography for Life: National Geography Standards.* Washington, D.C.: National Geographic Society, 1994.

Loomer, Scott A. MicroCAM for Windows Version 2.02. Available at wolf.its.ilstu.edu/microcam/index.html, 1999. World Wide Web.

Makower, Joel, ed. *The Map Catalog: Every Kind of Map and Chart on Earth and Even Some above It.* 3rd ed. New York: Vintage Books, 1992.

National Council for the Social Studies (NCSS). *Expectations of Excellence: Curriculum Standards for Social Studies.* Washington, D.C.: NCSS, 1994.

National Council of Teachers of Mathematics (NCTM). *Curriculum and Evaluation Standards for School Mathematics.* Reston, Va.: NCTM, 1989.

National Geographic Society (NGS). *National Geographic Atlas of the World.* 7th ed. Washington, D.C.: NGS, 1999.

Snyder, John P. *Map Projections: A Working Manual.* U.S. Geological Survey Professional Paper 1395. Washington, D.C.: United States Government Printing Office, 1987.

Sobol', Ilya M. *A Primer for the Monte Carlo Method.* Ann Arbor, Mich.: CRC Press, 1994.

Stoltman, Joseph P. "Research on Geography Teaching." In *Handbook of Research on Social Studies Teaching and Learning,* edited by James P. Shaver, pp. 437–47. New York: Macmillan Publishing Co., 1991.

Stout, William F., John Marden, and Kenneth J. Travers. *Statistics: Making Sense of Data.* 2nd ed. Rantoul, Ill.: Mobius Communications, 1999.

Travers, Kenneth J., William F. Stout, James H. Swift, and J. Sextro. *Using Statistics.* Menlo Park, Calif.: Addison Wesley Longman, 1985.

Wikle, Thomas. "Computer Software for Displaying Map Projections and Comparing Distortions." *Journal of Geography* 90 (November–December 1991): 264–67.

Wilkins, Jesse L. M. "The Cereal Box Problem Revisited." *School Science and Mathematics* 99 (3) (1999): 117–23.

ESTIMATING THE PERCENT OF THE EARTH THAT IS COVERED BY WATER   SHEET 1

1. Guess the percent of the earth that is covered by water.

2. With your group or class, discuss how you might estimate this ratio to check your guess.

The globe is the most accurate representation of the earth with regard to area, distance, direction, and shape. Therefore, a measure of the water coverage shown on a globe gives the best estimate of the percent of the earth that is covered with water. In this activity, you use a globe to estimate the percent of the earth that is covered by water. The results of this activity are used as a benchmark for comparing estimates of water coverage that are based on different map projections. Using a random device to produce random outcomes to investigate a given problem is known as a *Monte Carlo simulation*. A Monte Carlo simulation using the globe might proceed as follows:

1. Identify the model.	To estimate water coverage of the earth, we need a model of the earth. Let the inflatable globe represent the earth.
2. Define a trial.	To prepare for the trial, make an ink dot on your right index finger. A trial in this experiment consists of a toss of an inflatable globe around the room and an examination of whether the ink dot is in water or on land when the globe is caught.
3. Record statistics.	Announce to the class whether the ink-dotted finger is on land or in water, and record the data in the chart below.
4. Repeat the trial.	Decide on a prescribed number of trials to obtain enough data to calculate reasonable estimates. One hundred tosses and catches should be sufficient.
5. Calculate desired statistics.	Use the data to calculate the percent of water hits and the percent of land hits.

3. Use the following chart to keep a running tally of hits.

Land Hits	Water Hits

4. Use the data from the simulation to calculate the percents representing the number of times that the point landed in water or on land compared to the total number of trials.

Water: _____   Land: _____

5. What do these numbers estimate?

6. How do these results compare with your original guess for the percent of the earth that is covered by water?

From the *Mathematics Teacher*, November 2001

The chart below shows the estimated land area of the seven continents of the world (North America, South America, Australia, Africa, Antarctica, Asia, and Europe). Discuss with your group or class the relative size of the continents.

1. Which continent do you think has the largest land area? _____

2. Which continent do you think has the smallest land area? _____

3. Complete the chart by matching the seven continents with their appropriate estimated land area.

Continent	Land Area (Square Kilometers)
	30,065,000
	13,209,000
	7,687,000
	44,579,000
	9,938,000
	24,474,000
	17,819,000
Diameter of the earth	12,756 km

Source: National Geographic Society (NGS), *National Geographic Atlas of the World*, 7th ed. (Washington, D.C.: NGS, 1999)

4. Use the data in the chart to calculate the percent of the earth's surface that is covered by water. (Hint: the earth is approximately a sphere; the surface area of a sphere is $4\pi r2$.)

5. With your group or class, compare the results from question 4 with the results obtained from the simulation. Are they similar? If not, explain any discrepancies.

Figures 1, 2, and 3 present three projections of the globe onto a flat surface. All flat maps distort one or more aspects of their representation of the earth. Map projections themselves are based on the requirements of a given situation and the characteristics that are being investigated. Most map projections are constructed to maintain area, distance, direction, shape, or a combination of these characteristics. Some mapmakers are interested in showing accurate relationships of the areas of land and water masses. Other maps are concerned with accurate representations of distances between points on the surface of the earth. Others maintain accurate direction in terms of north, south, east, and west, whereas others try to make the shape of land and water areas accurately represent the actual shapes found on the earth's surface. In this activity, you investigate how different projections affect area. The following chart lists some properties of three projections and applications of the projections.

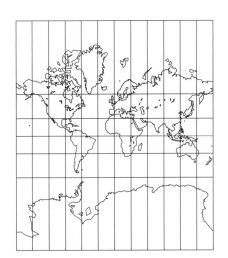

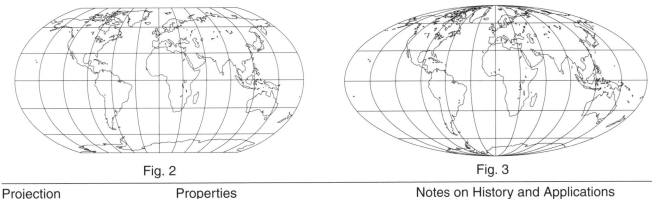

| Fig. 2 | Fig. 3 |

Projection	Properties	Notes on History and Applications
Mercator (fig. 1)	• The Mercator projection is not an equal-area map and greatly distorts size and shape the further that one moves toward the North and South Poles. For example, Greenland appears much larger than South America, although it is only one-eighth the size of South America on the globe • The Mercator projection is conformal, that is, the scale at any point on the map is the same in any direction. • Lines of longitude and latitude intersect at right angles. Angle measures are maintained.	The Mercator projection (1569) is probably the most well-known projection. Its use as a large-scale map of the world has been questioned by the National Geographic Society for its distortion of shape; for example, North America and Eurasia appear much larger than South America and Africa. It is useful for topographic maps and navigational charts because it maintains true angles.
Robinson (fig. 2)	• The Robinson projection is a compromise projection and not an equal-area map. • It distorts shape, scale, area, and distance to balance the error of projection properties. It presents the entire earth with reasonable shapes. • At high latitude, area is greatly distorted to allow better shapes in mid- and low-latitude regions.	The National Geographic Society used the Robinson projection (1963) as its official view of the world from 1988 to 1998. It was developed by Arthur Robinson, who directed the U.S. Office of Strategic Map Divisions during World War II, to overcome the problems that resulted from distortions when using such rectangular maps as the Mercator.
Mollweide (fig. 3)	The Mollweide projection is an equal-area map within an ellipse. That is, it maintains on the map the proportional relationships related to area.	The Mollweide projection (1805) is used for world maps and also for such large regions as the Pacific Ocean. It has also served as the inspiration for other equal-surface-area projections.

1. Use the chart to determine which map projection, if any, does the following:
   a) Underestimates the percent of water covering the earth; why?

   b) Overestimates the percent of water covering the earth; why?

   c) Provides a reasonable estimate of the percent of water covering the earth; why?

2. Using the map projections in figures 1, 2, and 3, discuss with your group or class how you might modify the globe simulation to estimate the percent of water covering the earth. Outline the steps for the simulation.

One way to carry out a similar simulation is to create a coordinate system for the flat maps, generate random numbers that represent points on the coordinate system, and record whether the point corresponds to a point in water or on land.

Use a transparency of graph paper to create a coordinate system that can overlay each map projection. On the graph paper, label an x- and y-axis that encompass both the horizontal and the vertical ranges of the maps. For example, label the x-axis 0–100 and the y-axis 0–50.

Using a calculator with a random-number generator or other random device, randomly generate a pair of integers within the assigned coordinate system. For example, the randInt function on the Texas Instruments calculators produces a random number between any two integers: randInt(0,100) produces an integer between 0 and 100, inclusive. A coordinate pair can be produced by using two randInt functions within brackets separated by a comma: {randInt(0,100),randInt(0,50)}. By continuing to press ENTER, you can continue to produce coordinate pairs. Generate at least fifty pairs, plot the points, and use the following chart to record whether they are on land or in water.

Projection	Land	Water	Percent Water
Mercator			
Mollweide			
Robinson			

3. How do your estimates for each projection compare with your findings on sheets 1 and 2?
   a) Which map projection gave an estimate closest to 70 percent? Why?

   b) Which map projection gave the worst estimate? Why?

4. Do these estimations match your expectations? Explain.

5. Share and discuss your estimates with the class. Combine data to form a class estimate for each map. How do your estimates compare with the class estimates?

1. On the Mercator map projection, what would the percent of water and the percent of land have been if all the points had fallen within—

   a) the top or bottom thirds of the map?

   b) the middle third of the map?

2. Redo the simulation; and keep track of whether the points fall in the northern third (N), the equatorial third (E), or the southern third (S) of your map. The map can be divided vertically using the x-coordinates of the graph paper. Use the following chart to record the hits and to   calculate percents.

Projection	Land	Water	Percent Water
N			
E			
S			

3. Did the estimates of water coverage seem to be affected by the distribution of dots over the map? Explain.

4. How could this knowledge help explain differing estimates from different groups in the class?

1. How do map projections that offer differing sizes, positions, and shapes of land masses affect our view of the world?

2. Consider the map projection below. How does this map differ from your perception of the world?

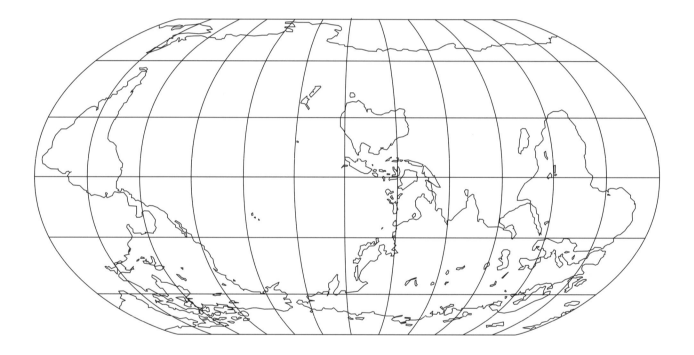

3. Why do maps always show either North America or Europe in the middle?

Sports and Distance-Rate-Time

Patrick R. Perdew

March 2002

Edited by Betty Krist, *University of Buffalo, New York*

Learning mathematics with understanding is essential" (NCTM 2000, p. 20). From their own experiences, students know the relationship between the speed of a ball and the time that a player has to react, and this knowledge can be used to facilitate students' understanding of uniform motion problems (Perdew 1999). Studying distance-rate-time relationships fulfills a Measurement Standard for grades 6–8 because it includes using derived measurements for rates (NCTM 2000). Converting speed units from the commonly used miles per hour to feet per second, which is more applicable in sports settings, also enhances students' understanding of the magnitude of such speeds. Expectations under the Measurement Standard for grades 9–12 include facility with unit analysis in keeping track of units during computations that involve converting units (NCTM 2000). Because only magnitude—and not direction—is important in this article, the teacher can emphasize the distinction between speed (a scalar quantity that refers to how fast an object is moving) and velocity (a vector quantity that refers to the rate at which an object changes its position).

TEACHER'S GUIDE

In many sports, the time that an athlete takes to react and make a decision is crucial. How quickly must Sammy Sosa decide whether to swing at a fastball? How long does a softball pitch from Olympian Danielle Henderson take to reach the plate? How much time does Anna Kournikova have to return a serve from Venus Williams? In this activity, students use unit analysis and the distance-rate-time formula to answer such questions.

The time an athlete takes to react and make a decision is crucial

Instructors may wish to introduce the subject with an example in which knowing speed in feet per second is more useful than knowing the number of miles per hour. The following example also uses the uniform motion formula.

Driver's manuals recommend a two-second following distance. The distance between vehicles should therefore equal the distance (in feet) that they travel in two seconds. To find this distance for cars going 60 miles per hour, we should first convert the rate to feet per second.

60 miles per hour

$$= \frac{60 \text{ hour}}{1 \text{ hour}}$$

$$= \frac{60 \text{ hour}}{1 \text{ hour}} \cdot \frac{5280 \text{ feet}}{1 \text{ mile}} \cdot \frac{1 \text{ hour}}{60 \text{ minutes}} \cdot \frac{1 \text{ minute}}{60 \text{ seconds}}$$

$$= \frac{88 \text{ feet}}{1 \text{ second}}$$

$$= 88 \text{ feet per second}$$

We next find the recommended following distance:

$$\text{distance} = \text{rate} \cdot \text{time}$$
$$= 988 \text{ feet per second}) \cdot (2 \text{ seconds})$$
$$= 176 \text{ feet}$$

The activities in this article involve finding the time when the distance and the rate are known. By solving the formula for time, instructors can demonstrate the way to rewrite the uniform motion formula:

$$d = r \cdot t$$
$$\frac{d}{r} = \frac{r \cdot t}{r}$$
$$\frac{d}{r} = t \ .$$

So

$$t = \frac{d}{r} \ .$$

The dimensions of a baseball field, a softball field, and a tennis court are discussed on the activity sheets so that students can determine the distance involved in each problem. Rates of speed are already provided. For further exercises, teachers may record and show the class a baseball game, softball game, or a tennis match or ask students to watch these sports independently. A teacher who decides to record an event should obtain permission

---

Patrick Perdew, teaches at Austin Peay State University, Clarksville, Tennessee. His interests include gender-equity issues and technology in the classroom.

from the appropriate network and sports association. Students can then write down the speeds of some of the pitches and serves and use the rates given by commentators or shown on screen instead of those given on the activity sheets to rework the exercises on the appropriate baseball, softball, or tennis activity sheet.

The speed of the ball is measured when it leaves the pitcher's hand or server's racket. So friction from air resistance, as well as the bounce of a fair serve before it reaches the returner's service line lessens the speed a little before the ball reaches the other player. This friction is generally negligible and is not examined in this article.

SOLUTIONS

Sheet 1

1. You should convert it to feet per second.
2. 5280 feet; 60 minutes; 60 seconds
3. The units in the numerator are feet; the units in the denominator are seconds.
4. 90 MPH

$$= \frac{90 \text{ miles}}{1 \text{ hour}}$$

$$= \frac{90 \text{ miles}}{1 \text{ hour}} \cdot \frac{5280 \text{ feet}}{1 \text{ mile}} \cdot \frac{1 \text{ hour}}{60 \text{ minutes}} \cdot \frac{1 \text{ minute}}{60 \text{ seconds}}$$

$$= \frac{475,200 \text{ feet}}{3600 \text{ seconds}}$$

$$= \frac{132 \text{ feet}}{1 \text{ second}}$$

$$= 132 \text{ feet per second.}$$

5. The rate of the baseball is 132 feet per second, and the distance that it travels is 60.5 feet. To the nearest thousandth of a second, the baseball will take

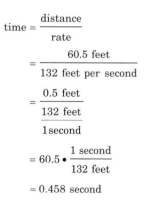

$$\text{time} = \frac{\text{distance}}{\text{rate}}$$

$$= \frac{60.5 \text{ feet}}{132 \text{ feet per second}}$$

$$= \frac{0.5 \text{ feet}}{\frac{132 \text{ feet}}{1 \text{ second}}}$$

$$= 60.5 \cdot \frac{1 \text{ second}}{132 \text{ feet}}$$

$$\approx 0.458 \text{ second}$$

to reach home plate.

Sheet 2

1. It should be converted to feet per second.
2. 65 MPH

$$= \frac{65 \text{ miles}}{1 \text{ hour}} \cdot \frac{5280 \text{ feet}}{1 \text{ mile}} \cdot \frac{1 \text{ hour}}{60 \text{ minutes}} \cdot \frac{1 \text{ minute}}{60 \text{ seconds}}$$

$$= \frac{65 \cdot 5280 \text{ feet}}{60 \cdot 60 \text{ seconds}}$$

$$= \frac{343,200 \text{ feet}}{3600 \text{ second}}$$

$$= 95 \, 1/3 \text{ feet per second.}$$

3. The rate of the softball is 95 1/3 feet per second, and the distance that it travels is 40 feet. Rounding to the nearest thousandth of a second, the ball will take

$$\text{time} = \frac{\text{distance}}{\text{rate}}$$

$$= \frac{40 \text{ feet}}{95 \, 1/3 \text{ feet per second}}$$

$$\approx 0.420 \text{ second}$$

to reach home plate.
4. The baseball took 0.458 second to reach home plate.
5. The softball batter must react more quickly. Answers may vary, but many students may have expected the baseball player to need a quicker reaction.

Sheet 3

1. 105 MPH

$$= \frac{105 \text{ miles}}{1 \text{ hour}} \cdot \frac{5280 \text{ feet}}{1 \text{ mile}} \cdot \frac{1 \text{ hour}}{60 \text{ minutes}} \cdot \frac{1 \text{ minute}}{60 \text{ seconds}}$$

$$= \frac{105 \cdot 5280 \text{ feet}}{60 \cdot 60 \text{ seconds}}$$

$$= \frac{554,400 \text{ feet}}{3600 \text{ second}}$$

$$= 154 \text{ feet per second.}$$

2. The rate of the tennis ball is 154 feet per second, and the distance that it travels is 78 feet. Rounding to the nearest thousandth of a second, the time needed to reach the returner is

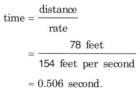

$$\text{time} = \frac{\text{distance}}{\text{rate}}$$

$$= \frac{78 \text{ feet}}{154 \text{ feet per second}}$$

$$\approx 0.506 \text{ second.}$$

3. The softball player must respond more quickly; answers will vary.

Sheet 4

1. The length of the segment between the center marks is 78 feet; the length of the segment from the center mark of the baseline to the location at which the ball would cross it is 17.55 feet. Rounded to the nearest hundredth of a foot,

$$c = \sqrt{a^2 + b^2}$$
$$= \sqrt{(78)^2 + (17.55)^2}$$
$$= \sqrt{6084 + 308.0025}$$
$$= \sqrt{6392}$$
$$= \sqrt{79.95} \; feet.$$

2. 120 MPH

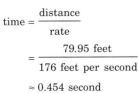

$$= \frac{120 \text{ miles}}{1 \text{ hour}} \cdot \frac{5280 \text{ feet}}{1 \text{ mile}} \cdot \frac{1 \text{ hour}}{60 \text{ minutes}} \cdot \frac{1 \text{ minute}}{60 \text{ seconds}}$$

$$= \frac{120 \cdot 5280 \text{ feet}}{60 \cdot 60 \text{ seconds}}$$

$$= \frac{633,600 \text{ feet}}{3600 \text{ second}}$$

$$= 176 \text{ feet per second.}$$

3. The rate of the tennis ball is 176 feet per second, and the distance that it travels is approximately 79.95 feet. Rounded to the nearest thousandth of a second, it reaches the returner in

$$\text{time} = \frac{\text{distance}}{\text{rate}}$$
$$= \frac{79.95 \text{ feet}}{176 \text{ feet per second}}$$
$$\approx 0.454 \text{ second}$$

4. Andre Agassi must react more quickly to the ball. Other answers will vary; some students may have expected Anna Kournikova because they may have thought that cross court is so much farther that the 15 MPH difference in the serves would not compensate for it.

5. Rounded to the nearest thousandth of a second,

$$\text{time} = \frac{\text{distance}}{\text{rate}}$$
$$= \frac{78 \text{ feet}}{176 \text{ feet per second}}$$
$$\approx 0.443 \text{ second.}$$

6. Softball (0.420 second) requires the athlete to react the fastest.

REFERENCES

National Council of Teachers of Mathematics (NCTM). *Principles and Standards for School Mathematics.* Reston, Va.: NCTM, 2000.

Perdew, Patrick. "Keep Your Eye on the Ball: Uniform Motion and Reaction Time." *The Math Projects Journal* 2 (March–April 1999): 11.

On a baseball diamond, the distance from the pitcher's mound to home plate is 60 feet 6 inches, or 60.5 feet. If Tom Glavine throws Sammy Sosa a 90 MPH fastball, how quickly does it reach home plate?

1. The rate is given in miles per hour. Into what units should you convert it? _____

2. To convert it to feet per second, fill in the blank beside each unit with the appropriate number.

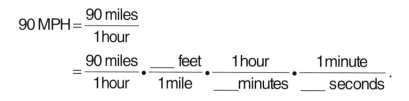

$$90\,\text{MPH} = \frac{90\,\text{miles}}{1\,\text{hour}}$$

$$= \frac{90\,\text{miles}}{1\,\text{hour}} \cdot \frac{\_\_\_\ \text{feet}}{1\,\text{mile}} \cdot \frac{1\,\text{hour}}{\_\_\_\text{minutes}} \cdot \frac{1\,\text{minute}}{\_\_\_\text{seconds}} \cdot$$

3. Use unit analysis to divide out common units. What units are now in the numerator?

_____

What units are in the denominator? _____

4. Fill in the blanks, and complete the conversion in problem 2:

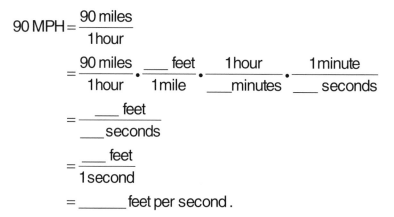

$$90\,\text{MPH} = \frac{90\,\text{miles}}{1\,\text{hour}}$$

$$= \frac{90\,\text{miles}}{1\,\text{hour}} \cdot \frac{\_\_\_\ \text{feet}}{1\,\text{mile}} \cdot \frac{1\,\text{hour}}{\_\_\_\text{minutes}} \cdot \frac{1\,\text{minute}}{\_\_\_\ \text{seconds}}$$

$$= \frac{\_\_\_\ \text{feet}}{\_\_\_\text{seconds}}$$

$$= \frac{\_\_\_\ \text{feet}}{1\,\text{second}}$$

$$= _____\ \text{feet per second}\,.$$

5. The rate of the baseball is _____ feet per second, and the distance that it travels is _____. Rounding to the nearest thousandth of a second, in how much time does the baseball reach home plate?

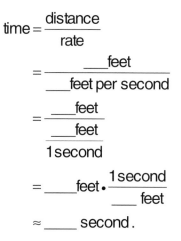

$$\text{time} = \frac{\text{distance}}{\text{rate}}$$

$$= \frac{\_\_\_\text{feet}}{\_\_\_\text{feet per second}}$$

$$= \frac{\_\_\_\text{feet}}{\dfrac{\_\_\_\text{feet}}{1\,\text{second}}}$$

$$= \_\_\_\text{feet} \cdot \frac{1\,\text{second}}{\_\_\_\ \text{feet}}$$

$$\approx \_\_\_\ \text{second}\,.$$

Under Olympic rules, the distance from the pitcher's mound to home plate is 40 feet on a softball diamond. If Olympic softball pitcher Lisa Fernandez hurls a 65 mph fastball, how quickly does it cross home plate?

1. The rate of the softball, like that of the baseball, should be converted to _____.

2. To convert, fill in the blank beside each unit with the appropriate number. Use unit analysis to divide out common units and to perform the arithmetic operations.

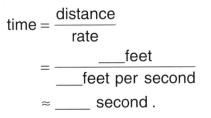

$$65\,\mathrm{MPH} = \frac{65\ \text{miles}}{1\,\text{hour}} \cdot \frac{\underline{\quad}\ \text{feet}}{1\,\text{mile}} \cdot \frac{1\,\text{hour}}{\underline{\quad}\ \text{minutes}} \cdot \frac{1\,\text{minute}}{\underline{\quad}\ \text{seconds}}$$

$$= \frac{65 \cdot \underline{\quad}\ \text{feet}}{\underline{\quad} \cdot \underline{\quad}\ \text{seconds}}$$

$$= \frac{\underline{\quad}\ \text{feet}}{\underline{\quad}\ \text{seconds}}$$

$$= \underline{\qquad}\ \text{feet per second}.$$

3. The rate of the softball is _____ feet per second, and the distance that it travels is _____ feet. Rounding to the nearest thousandth of a second, in how much time does it reach home plate?

$$\text{time} = \frac{\text{distance}}{\text{rate}}$$

$$= \frac{\underline{\quad}\,\text{feet}}{\underline{\quad}\,\text{feet per second}}$$

$$\approx \underline{\quad}\ \text{second}.$$

4. In problem 5 of sheet 1, how much time did the baseball take to reach home plate?

5. Compare the time for the baseball with the time that the softball took to reach the plate. Which batter (softball or baseball) must react more quickly, that is, in less time, to the pitch?

Did you expect this answer? Why or why not?

From baseline to baseline, a tennis court is 78 feet long. A tennis ball must therefore travel 78 feet between the server and returner if the server stands at the center mark of the baseline and serves down the opponent's half-court line while the returner waits for the ball at his or her own baseline. Otherwise, since the serve must go cross court, the actual distance of the serve is the hypotenuse of a right triangle. How long will a 105 MPH down-the-line serve from Venus Williams take to reach Anna Kournikova?

　　1. Convert the rate to feet per second.

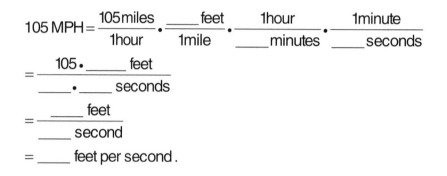

$$105\,\text{MPH} = \frac{105\,\text{miles}}{1\,\text{hour}} \cdot \frac{\underline{\quad}\,\text{feet}}{1\,\text{mile}} \cdot \frac{1\,\text{hour}}{\underline{\quad}\,\text{minutes}} \cdot \frac{1\,\text{minute}}{\underline{\quad}\,\text{seconds}}$$

$$= \frac{105 \cdot \underline{\quad}\,\text{feet}}{\underline{\quad} \cdot \underline{\quad}\,\text{seconds}}$$

$$= \frac{\underline{\quad}\,\text{feet}}{\underline{\quad}\,\text{second}}$$

$$= \underline{\quad}\ \text{feet per second}.$$

　　2. The rate of the tennis ball is _____ feet per second, and the distance that it travels is _____ feet. Rounding to the nearest thousandth of a second, in how much time does the ball reach the returner?

$$\text{time} = \frac{\text{distance}}{\text{rate}}$$

$$= \frac{\underline{\quad}\,\text{feet}}{\underline{\quad}\,\text{feet per second}}$$

$$\approx \underline{\quad}\ \text{second}.$$

　　3. Compare this answer with the times that the baseball and softball took to reach the plate on sheets 1 and 2. Which player must react more quickly to the ball?

Did you expect this result? Why or why not?

If Pete Sampras stands at his center mark and hits a 120 MPH serve cross court to the corner formed by his opponent's baseline and sideline, how much time does Andre Agassi have to return it?

Again, from the center mark of one baseline to the center mark of the other baseline is 78 feet, but the ball must bounce in the service box, which is 60 feet from the server. A line segment connecting the center mark of the server's baseline with the center mark of the receiver's service box is 60 feet long. The width of the fair area in the tennis court, from sideline to sideline, is 27 feet. A line segment drawn from the center mark of the service box to the corner where the end of the baseline meets the sideline is 13.5 feet long. The line segments form the legs of a right triangle, with the center mark at the vertex.

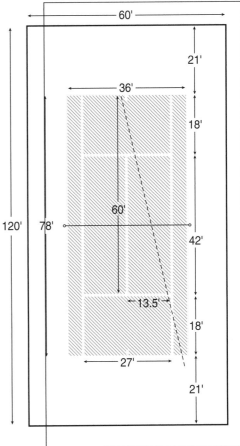

After bouncing in the service box, the ball continues in the direction of the baseline. Two similar triangles are formed. The legs of one are 60 feet and 13.5 feet; the legs of the other are 78 feet and x feet. Solving the similar triangles, we get

$$\frac{60}{13.5} = \frac{78}{x}$$

or $x = 17.55$ feet. Using the Pythagorean theorem, $c^2 = a^2 + b^2$, the distance that the tennis ball traveled from Sampras's center mark to the corner of Agassi's baseline and sideline can be found as the length of a hypotenuse.

1. Let a equal the length of the segment between center marks, that is, the length of the court. This length is _____ feet. Let b equal the length of the segment from the center mark of the baseline to the location at which the ball would cross it. This length is _____ feet. Rounded to the nearest hundredth of a foot,

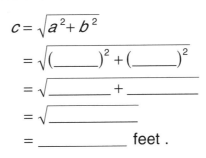

$$c = \sqrt{a^2 + b^2}$$
$$= \sqrt{(\underline{\quad})^2 + (\underline{\quad})^2}$$
$$= \sqrt{\underline{\quad\quad} + \underline{\quad\quad}}$$
$$= \sqrt{\underline{\quad\quad}}$$
$$= \underline{\quad\quad} \text{ feet .}$$

2. Convert the rate to feet per second.

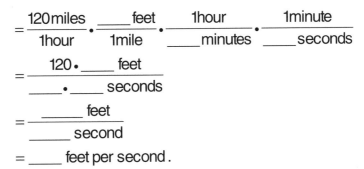

$$= \frac{120\,\text{miles}}{1\,\text{hour}} \cdot \frac{\underline{\quad}\,\text{feet}}{1\,\text{mile}} \cdot \frac{1\,\text{hour}}{\underline{\quad}\,\text{minutes}} \cdot \frac{1\,\text{minute}}{\underline{\quad}\,\text{seconds}}$$
$$= \frac{120 \cdot \underline{\quad}\,\text{feet}}{\underline{\quad} \cdot \underline{\quad}\,\text{seconds}}$$
$$= \frac{\underline{\quad}\,\text{feet}}{\underline{\quad}\,\text{second}}$$
$$= \underline{\quad}\,\text{feet per second .}$$

3. The rate of the tennis ball is _____ feet per second, and the distance that it travels is approximately _____ feet. Rounding to the nearest thousandth of a second, in how much time does it reach the returner?

$$\text{time} = \frac{\text{distance}}{\text{rate}}$$

$$= \frac{\underline{\quad\quad} \text{ feet}}{\underline{\quad\quad} \text{ feet per second}}$$

$$\approx \underline{\quad\quad} \text{ second}.$$

4. Compare this result with the time obtained for the tennis ball to reach the returner in sheet 3. Which player must react more quickly to the ball? _____

   Did you expect this result? Why or why not?

5. If Sampras had served down the half-court line at 120 MPH instead of cross-court, how much time would the ball take to reach Agassi? (Proceed as in sheet 3.)

6. Among all the sports discussed in this activity, which one requires the athlete to react most quickly, that is, in the least amount of time, to the ball?

Print-Shop Paper Cutting: Ratios in Algebra

Careylyn Hill

April 2002

Edited by Betty Krist, *University at Buffalo, Buffalo, New York*

> A major goal of high school mathematics is to equip students with knowledge and tools that enable them to formulate, approach, and solve problems beyond those that they have studied. High school students should have significant opportunities to develop a broad repertoire of problem-solving (or heuristic) strategies. They should have opportunities to formulate and refine problems because problems that occur in real settings do not often arrive neatly packaged. Students need experience in identifying problems and articulating them clearly enough to determine when they have arrived at solutions. The curriculum should include problems for which students know the goal to be achieved but for which they need to specify—or perhaps gather from other sources—the kinds of information needed to achieve it.
>
> —*Principles and Standards for School Mathematics*

Students have a difficult time transferring mathematics learned from a textbook to an activity that uses real data or that simulates a real-world experience. Several years ago, a vocational teacher expressed his frustration that students were not able to do the activities in a paper-cutting chapter in the course that he taught. In response, I borrowed the textbook (Cogoli 1980) and wrote the following activity for algebra students. This paper-cutting activity was developed specifically to help students gain experience in transferring skills and information that they have learned to different situations. The content focus of the activity is using ratios for various jobs that an employee in a print shop might encounter in a day's work.

Objectives. This mathematics activity helps students see connections between mathematics in the classroom and mathematics in the real world. Specifically, students learn to set up and solve ratios in a new way. Students also learn how to deal with remainders in a nontraditional way. Depending on the application, students truncate ratios or round them up. In addition, students interpret scale drawings and create an accurate scale drawing.

Prerequisites. Students should be familiar with ratios, percents, and scale drawings.

Grade levels. 7–12

Materials. Activity sheets, twelve 5-by-7-inch sheets of paper per student group, 17-by-22-inch stock sheets for each group, rulers, tape, and staplers; optional supplies include a ream of paper for demonstration purposes, 22 1/2-by-28 1/2-inch colored posterboard, and six 8 1/2-by-11-inch sheets of white paper.

Time required. 45–60 minutes

DIRECTIONS

Students can do this activity individually, or they can work in groups of three or four students. During the brief presentation that I use to introduce the activity, I inform students that this activity allows them to practice a skill needed when working in a print shop and gives them practice in doing some out-of-the-ordinary mathematical computations. I tell them that they can master the skills by using the material that they have learned in class, by doing some thinking, and by using problem-solving strategies. To help with vocabulary, I show the class a ream of paper and set it on the chalk tray. I also put a sheet of colored posterboard, labeled "stock sheet," on the tray. I then have 8 1/2-by-11-inch sheets of paper labeled "press sheets" to show students options for cutting press sheets from a stock sheet.

Student instructions

The student instructions give information, formulas, diagrams, and an example of a print-shop application. The students need this information to solve problems in the activity. They should read the material carefully and ask other students or the instructor to help clarify any part that they do not understand.

Careylyn Hill taught most recently at Canyon Del Oro High School, Tucson, Arizona. She wrote this article while finishing her master's degree at the University of Arizona, Tucson, Arizona, and is currently a stay-at-home mother of three.

Sheet 1

The first two questions are factual questions about the reading material. These questions help students begin doing the ratio computations themselves. Question 1 lends itself toward a discussion of when they must compute both ratios and when they need to compute only one ratio. Question 3 asks students to connect the skill that they use in this activity with a skill that they might need.

Sheet 2

Job 1. This problem takes the students through the preliminary work for a job at a print shop. This job draws on ideas learned in the student instructions.

Job 2. This activity extends job 1 by asking students to consider the cost of paper.

Job 3. This job takes the activity to a level of greater complexity. The students make an accurate scale drawing of the layout of press sheets on a 17-by-22-inch stock sheet of paper. If this size is not available, the teacher should determine an appropriate scale and use paper that is available. Using the same size sheet, the students staple or tape the precut 5-by-7-inch sheets to learn whether they can find a more efficient way to line up press sheets than the number that they computed by using ratios and their drawing.

When the students are done with the activity, we discuss it. I ask students when a more efficient way might exist to cut the stock paper than the results indicated by the ratios. If the remainders are large, a more efficient result might be possible. Students can line up the ten press sheets in three different ways. I show the various ways on the board. See figures 1a, 1b, and 1c. I ask students which of the three layouts they would choose if they were actually cutting the press sheets. Then I ask the students why they selected that arrangement. Figure 1a and 1c are simpler to measure, draw, and cut than figure 1b.

SOLUTIONS

Sheet 1

1.

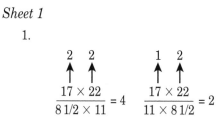

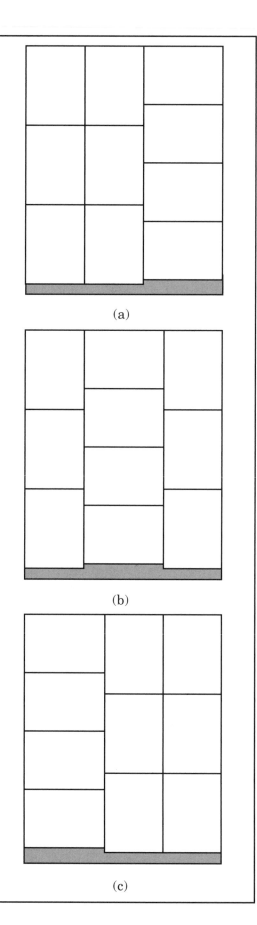

(a)

(b)

(c)

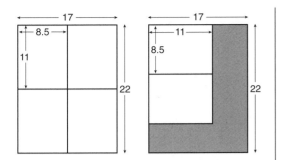

2. The number of letterheads that can be cut from two reams of 17-by-22-inch stock is $(4 \times 500 \times 2)$, or 4000.

3. Answers will vary.

Sheet 2

1. Job 1

$$\frac{\overset{2}{\uparrow}\ \overset{2}{\uparrow}}{17 \times 22}{8\ 1/2 \times 11} = 4 \qquad \frac{\overset{1}{\uparrow}\ \overset{2}{\uparrow}}{17 \times 22}{11 \times 8\ 1/2} = 2$$

The number of stock sheets required for the job is twenty-eight. The diagram should be the same as the one for sheet 1, problem 1.

2. Job 2

$$\frac{\overset{3}{\uparrow}\ \overset{2}{\uparrow}}{17 \times 22}{5\ 1/2 \times 8} = 6 \qquad \frac{\overset{2}{\uparrow}\ \overset{4}{\uparrow}}{17 \times 22}{5\ 1/2 \times 8} = 8$$

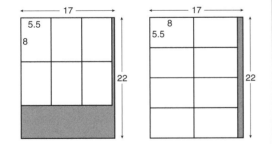

a) You need 1500 stock sheets.

b) You need 3 reams of stock paper.

c) The cost of the stock paper is $61.50.

3. Job 3

$$\frac{\overset{3}{\uparrow}\ \overset{3}{\uparrow}}{17 \times 22}{5 \times 7} = 9 \qquad \frac{\overset{2}{\uparrow}\ \overset{4}{\uparrow}}{17 \times 22}{7 \times 5} = 8$$

You need sixty-three stock sheets.

a) See figure 2.

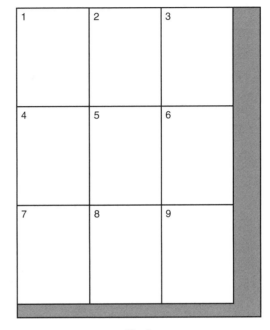

Fig. 2

b) Answers will vary; see figure 1 for some possible answers.

REFERENCES

Cogoli, John E. *Photo Offset Fundamentals*. Woodland Hills, Calif.: Glencoe/ McGraw-Hill, 1980.

National Council of Teachers of Mathematics (NCTM). *Principles and Standards for School Mathematics*. Reston, Va.: NCTM, 2000.

PAPER-CUTTING ACTIVITY: STUDENT INSTRUCTIONS

In this activity, you work in a print shop. The day's work includes several jobs, each of which first requires cutting sheets of paper to the proper size.

You need the following information before you can begin work.

A ream consists of 500 sheets of paper. Common papers are often packaged in this quantity.

To determine the number of 6-by-10-inch press sheets that can be obtained from a 22 1/2-by-28 1/2-inch stock sheet, calculate first as indicated in solution A, then change the orientation of the press sheets and calculate as indicated in solution B.

Divide the press-sheet size into the stock-sheet size. In solution B, switch the bottom numbers of your ratio.

Basic formula:

$$\frac{\text{Dimensions of the stock sheet} = \text{width} \cdot \text{height}}{\text{Dimensions of the press sheet} = \text{width} \cdot \text{height}}$$

Solution A:

$$\frac{22\frac{1}{2} \text{ inches wide} \times 28\frac{1}{2} \text{ inches high}}{6 \text{ inches wide} \times 10 \text{ inches high}}$$

Solution B:

$$\frac{22\frac{1}{2} \text{ inches wide} \times 28\frac{1}{2} \text{ inches high}}{10 \text{ inches wide} \times 6 \text{ inches high}}$$

Divide the denominator into the numerator for each dimension. Round down to the nearest whole number.

A:
$$\frac{\overset{3}{\underset{\uparrow}{}} \quad \bullet \quad \overset{2}{\underset{\uparrow}{}} \quad = 6}{\frac{22\frac{1}{2} \text{ inches wide} \times 28\frac{1}{2} \text{ inches high}}{6 \text{ inches wide} \times 10 \text{ inches high}}}$$

B:
$$\frac{\overset{2}{\underset{\uparrow}{}} \quad \bullet \quad \overset{4}{\underset{\uparrow}{}} \quad = 8}{\frac{22\frac{1}{2} \text{ inches wide} \times 28\frac{1}{2} \text{ inches high}}{10 \text{ inches wide} \times 6 \text{ inches high}}}$$

The following diagrams are visual representations of the two solutions:

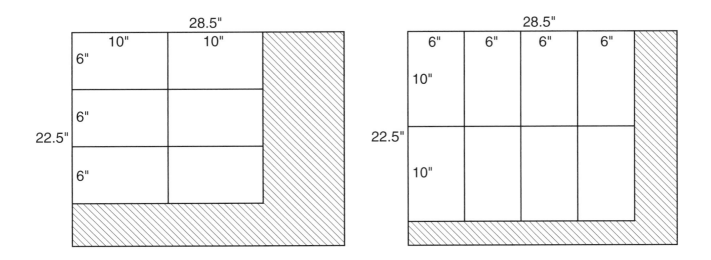

Solution B results in eight pieces, whereas solution A results in only six pieces. Because solution B wastes less paper than solution A, solution B is more efficient. However, you may be able to find a better solution. If you obtain large remainders when you perform the division, as in solutions A and B, you may be able to draw press sheets directly on the stock sheet and obtain a larger number of pieces per sheet. Refer to the diagram below.

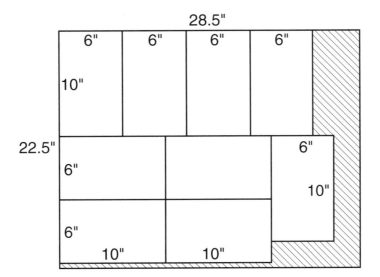

Diagram method of figuring stock

The diagram shows that nine 6-by-10-inch press sheets can be obtained from the same 22 1/2-by-28 1/2-inch stock sheet—one more press sheet per stock sheet than in solution B. Considerable savings could result if many sheets were to be cut.

1. How many 8 1/2-by-11-inch letterheads can be cut from a 17-by-22-inch stock sheet? Show both ratios, and sketch the diagrams. Label the diagrams.

2. Use the larger number of letterheads that can be cut from a stock sheet in question 1, and determine the number of 8 1/2-by-11-inch letterheads that can be cut from two reams of 17-by-22-inch stock.

3. Make a connection between the paper-cutting skill used in this activity and some situation or occupation that uses this skill.

1. Job 1

   The order asks for one hundred 8 1/2-by-11-inch letterheads. The stock size is 17 inches by 22 inches. Use a 10 percent allowance for spoilage, that is, sheets ruined during printing. How many stock sheets of paper are required for the job? You must write ratios or draw diagrams.

2. Job 2

   This order asks for 12,000 pieces of notepaper, each measuring 5 1/2 inches by 8 inches. The stock size is 17 inches by 22 inches, and it sells for $20.50 a ream. Assume that no spoilage occurs. Write ratios, and draw diagrams.

   a) How many stock sheets do you need? _____

   b) How many reams of stock paper do you need? _____

   c) What is the cost of the stock paper? _____

3. Job 3

   You have reached the last job of the day. However, it is a big one, so take your time and get it right.

   The last order requests 500 5-by-7-inch report cards. How many stock sheets of paper do you need if the stock-sheet size is 17 inches by 22 inches? Allow 10 percent for spoilage.

   Use ratios to show the two options that indicate the number of 5-by-7-inch sheets that you can get from one stock sheet.

   a) Use a 17-by-22-inch stock sheet. Complete the diagram by showing how you would cut the paper to get the most efficient result from your ratios. Use a ruler. Shade in the waste.

   b) You again have a 17-by-22-inch stock sheet. For this diagram, use the cutout 5-by-7-inch sheets of paper. Determine whether your previous diagram is in fact the most efficient way to cut the stock sheet. Whatever way you decide to cut the stock sheet, place the 5-by-7-inch sheets in that order, and staple or tape them to the diagram.